Badger GCSE Science

How Science Works

Chemistry

Andrew Grevatt
Dr. Deborah Shah-Smith

Badger
Publishing

INTRODUCTION

This is the first of a series of three books of 'How Science Works' activities for GCSE science. These have been developed by teachers to give them a range of resources to use in teaching science through the How Science Works principles. These activities work well as an 'add on' to any of the KS4 schemes of work.

How Science Works is now an essential part of the Science National Curriculum. For many science teachers, this has meant getting to grips with a new approach to science teaching. These activities have been developed to encourage structured discussion, improve knowledge and understanding of How Science Works and support learners to consider a range of viewpoints and make an informed decision.

The new KS4 Science National Curriculum sets out the parameters for How Science Works and the specifications for each of the GCSE examination boards have interpreted these in a variety of ways. We have developed these activities to be suitable for use in all schools, whatever exam board they use.

We have found that learners do not necessarily have the background knowledge about scientific issues such as genetically modified food. Without this background information, it is hard for them to discuss the issue and form an opinion. These How Science Works activities have been developed to help boost background knowledge of scientific issues, consider evidence from a range of contexts and support learners to make a decision.

HOW TO USE THESE ACTIVITIES

The general approach to using these activities is to introduce the learners to the task, then allow them to discuss the task in pairs or groups of four. During this time, the teacher should circulate amongst the groups to encourage discussion. Once the learners have had time to discuss their ideas and make decisions, the teacher can lead a discussion to draw out the key points.

More specifically, the activities take a variety of forms that require slightly different management. **Data analysis, graph drawing and predicting activities** follow the general approach but note that the discussion of the activities should focus on the *process* of analysing, data presentation and prediction rather than just what the answer is.

Card sorts take a variety of forms, but the general approach allows learners to physically sort statements into groups. Discussion should focus on the nature of evidence or facts and opinions in the context of the content of the activity.

Timelines require more time and can be extended by allowing learners to add their own images and additional events on the timeline. They are designed so that learners can identify how ideas change and relate images to the text. Follow-up discussion should focus on how and why the ideas have changed and perhaps imagining what may happen in the future.

A NOTE ABOUT TIMING

We have identified the approximate time taken for the activities based on an average ability class. More able groups may need less time on the task but more time on the discussion and lower ability learners may need support (particularly literacy) through the activities themselves. Most tasks have some differentiation suggestions.

AUTHORS

Andrew Grevatt is an experienced Advanced Skills Teacher who is currently an associate tutor at the University of Sussex, where he is researching for a professional doctorate in education.

Dr. Deborah Shah-Smith is an experienced science teacher. She has a keen interest in developing resources using practice-based evidence. She would like to dedicate this publication to her husband, Paul, and daughter, Zaveri.

ACKNOWLEDGEMENTS

We would like to thank our colleagues who have helped us to develop these tasks. These include Ben Riley of Oriel High School, West Sussex, Ross Palmer of Cardinal Newman School, Brighton and Hove, and the team at Badger Publishing for their hard work and patience.

CONTENTS AND CURRICULUM LINKS

CHEMICAL REACTIONS	NC HSW		ACTIVITY TYPE
1. The Fastest Fizz	1a	Data, evidence, theories and explanations	Methods of measurement
2. Fermentation	1a	Data, evidence, theories and explanations	Graph analysis and predictions
3. Which Metal is the Most Reactive?	2ac	Practical and enquiry skills	Planning an investigation
4. Collecting Carbon Dioxide	2d	Practical and enquiry skills	Planning an investigation
5. Evaluating Copper Extraction	2d	Practical and enquiry skills	Evaluating an investigation
6. Hazard Labels in the Laboratory	3ac	Communication skills	Interpretation
7. Forensic Testing	3ab	Communication skills	Interpretation
8. Brewing Beer	3a	Communication skills	Knowledge: card sort
9. Haber: Hero or Horrid?	4c	Applications and implications of science	Ethical issues: discussion and card sort
10. Metals Discovery Timeline	4c	Applications and implications of science	How ideas change: card sequencing

CHEMICAL PATTERNS			
11. Mendeleev's Predictions	1bc	Data, evidence, theories and explanations	Predicting
12. Element Sort	1ab	Data, evidence, theories and explanations	Discussion and card sort
13. Making Molar Solutions	2cd	Practical and enquiry skills	Discussion and card sort
14. The Size of Atoms	3abc	Communication skills	Graph drawing and analysis
15. Element Melting Point Patterns	3ac	Communication skills	Graph drawing and interpretation
16. Carbon Chains and Boiling Points	3abc	Communication skills	Graph drawing and predicting
17. Matching Molecules	3c	Communication skills	Card matching
18. Energy Changing Chemical Reactions	3abc	Communication skills	Data analysis and application of knowledge
19. Mobile Metals: Benefits, Drawbacks and Risks	4a	Applications and implications of science	Discussion and card sort
20. Structure of the Atom Timeline	4c	Applications and implications of science	How ideas change: card sequencing

NATURAL RESOURCES			
21. Why Recycle?	1ab	Data, evidence, theories and explanations	Graph interpretation
22. Making Biofuels	2a	Practical and enquiry skills	Planning an investigation, card sequencing
23. Fuel Economy	2b	Practical and enquiry skills	Text to table
24. Researching Limestone	2abcd	Practical and enquiry skills	Planning an investigation
25. Waste Tyre Graphs	3abc	Communication skills	Graph drawing
26. Fractional Distillation of Crude Oil	3ac	Communication skills	Graph drawing

NATURAL RESOURCES	NC HSW		ACTIVITY TYPE
27. World Oil Reserves: The Future?	3a	Communication skills	Graph interpretation
28. Desalination of Sea Water: Benefits, Drawbacks and Risks	4a	Applications and implications of science	Discussion and card sort
29. Desalination of Sea Water in Australia: Social, Economic and Environmental Issues	4b	Applications and implications of science	Discussion and card sort
30. Quarrying: Social, Economic and Environmental Issues	4b	Applications and implications of science	Discussion and card sort

MATERIAL PROPERTIES			
31. Artificial Fertilisers	1ab	Data, evidence, theories and explanations	Graph analysis and predictions
32. Do Food Colourings Affect Behaviour?	1d	Data, evidence, theories and explanations	Evidence evaluation
33. Which Washing Powder Washes Cleanest?	2ac	Practical and enquiry skills	Planning an investigation
34. Thermoplastic or Thermosetting Plastic?	2a	Practical and enquiry skills	Interpretation
35. Which Plastic Carrier Bag is Best?	3abc	Communication skills	Evidence evaluation
36. Scale: Big to Very Small	3c	Communication skills	Discussion and card sort
37. Washing Powder Properties: Essential or Desirable	4ab	Applications and implications of science	Discussion and card sort
38. Nanotechnology: Benefits, Drawbacks and Risks	4a	Applications and implications of science	Discussion and card sort
39. Artificial Fertilisers: Social, Economic and Environmental Issues	4b	Applications and implications of science	Discussion and card sort
40. Plastic Bags: Benefits, Drawbacks and Risks	4a	Applications and implications of science	Discussion and card sort

EARTH'S ATMOSPHERE			
41. Why did the Oxidation Catastrophe Happen?	1abcd	Data, evidence, theories and explanations	Comparing theories
42. History of the Earth's Atmosphere	1bcd	Data, evidence, theories and explanations	Discussion and card sort
43. Predicting Carbon Dioxide Levels	1bd	Data, evidence, theories and explanations	Data analysis and application
44. Greenhouse Gases: Which is Worst?	1abc	Data, evidence, theories and explanations	Presenting in tables
45. Acid Rain Research	2abd	Practical and enquiry skills	Planning using secondary resources
46. Planetary Atmospheres	3abc	Communication skills	Graph drawing and analysis
47. Fertiliser From the Air	3a	Communication skills	Card sort and sequencing
48. Is the Ozone Layer Recovering?	3ac	Communication skills	Graph drawing
49. Coal as an Energy Resource: Benefits, Drawbacks and Risks	4a	Applications and implications of science	Discussion and card sort
50. Growing Biofuels in Malawi: Social, Economic and Environmental Issues	4b	Applications and implications of science	Discussion and card sort

CHEMICAL REACTIONS
THE FASTEST FIZZ

1

KS4 NATIONAL CURRICULUM HSW LINK

1. *Data, evidence, theories and explanations*
   a. how scientific data can be collected and analysed

RESOURCES:
Task Sheet 1, enough for one per learner or one between two.

TIME:
10 minutes for the activity, 5 minutes for class discussion.

NOTES

- This task is most suitable for use as a Starter or Plenary.
- Teachers may want to show the equipment set-up for this investigation.
- Learners must be aware of these key words/concepts before attempting the task: measuring rates of reaction; dependent and independent variables.

SUGGESTED ANSWERS

A. i) Measure the change in mass with time.
   Advantages: the reaction can be tracked every few seconds.
   Disadvantages: some spray can be lost as well as gas.
   Human error with using a stop-watch.

   ii) Measure the total time taken for the reaction to finish.
   Advantages: it is quick to do.
   Disadvantages: it is difficult to tell when the reaction has finished.
   Human error with using a stop-watch.

B. i) Measure the change in mass with time.
   Independent variable: size of calcium carbonate pieces.
   Dependent variable: mass of reactants.

   ii) Measure the total time taken for the reaction to finish.
   Independent variable: size of calcium carbonate pieces.
   Dependent variable: time taken for reaction to finish.

C. Suggest ways to present the data using your chosen method.
   Line graph with line of best fit.

EXTENSION SUGGESTION

How could the investigation be made reliable and accurate?

THE FASTEST FIZZ

Isobel was investigating the effect of the size of calcium carbonate pieces on the rate of reaction with hydrochloric acid.

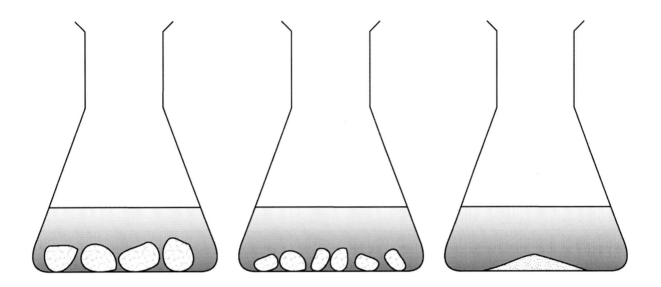

TASK

Isobel couldn't decide whether to:
  i)   measure the change in mass with time *or*
  ii) the total time taken for the reaction to finish.

A. What are the advantages and disadvantages of each method?

B. For your chosen method, identify the independent and dependent variables.

C. Suggest ways to present the data using your chosen method.

CHEMICAL REACTIONS
FERMENTATION

2

KS4 NATIONAL CURRICULUM HSW link

1. *Data, evidence, theories and explanations*
   a. how scientific data can be collected and analysed

RESOURCES:
Task Sheet 2, enough for one per learner or one between two.

TIME:
10 minutes for the activity, 5 minutes for class discussion.

NOTES

- This task is most suitable for use as a Starter, Main Activity or Plenary.
- Teachers may need to discuss the graph with lower ability learners before they attempt the task.
- Learners must be aware of these key words/concepts before attempting the task: graph interpretation; the process of fermentation; the fermentation reaction.

SUGGESTED ANSWERS

A. Identify the independent and dependent variables.
   Independent variable: time.
   Dependent variable: volume of carbon dioxide.

B. What units could you use to measure the volume of carbon dioxide given off?
   cm^3

C. Suggest why the volume of carbon dioxide given off remains constant at the end of the experiment.
   The reaction has stopped because the sugar has been used up during the fermentation reaction.

D. Describe in words what the graph shows is happening to the amount of carbon dioxide produced.
   Initially, carbon dioxide is given off slowly but, as the reaction progresses, the amount given off increases before reaching a steady amount per unit time. After this time, the amount of carbon dioxide given off slows until it stops being given off and the reaction stops.

EXTENSION SUGGESTION

Suggest ways to speed up the reaction.

This graph shows the volume of gas given off during a fermentation reaction.

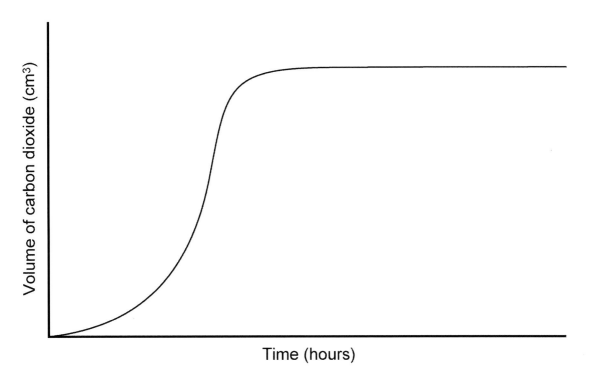

TASK

A. Identify the independent and dependent variables.

B. What units could you use to measure the volume of carbon dioxide given off?

C. Suggest why the volume of carbon dioxide given off remains constant at the end of the experiment.

D. Describe in words what the graph shows is happening to the amount of carbon dioxide produced.

CHEMICAL REACTIONS
WHICH METAL IS THE MOST REACTIVE?

3

KS4 NATIONAL CURRICULUM HSW LINK

2. *Practical and enquiry skills*
   a. plan to test a scientific idea, answer a scientific question or solve a scientific problem
   c. work accurately and safely, individually and with others, when collecting first-hand data

RESOURCES:
Task Sheet 3, enough for one between two learners.

TIME:
15 minutes for the activity, 10 minutes for class discussion.

NOTES

- This task is most suitable for use as an extended Starter, Main Activity or Plenary. It could also be set as a Homework Activity.
- Learners must be aware of these key words/concepts before attempting the task: planning an investigation; the reaction between a metal and an acid (hydrochloric acid).

SUGGESTED ANSWERS

Prediction with a scientific reason:
Lithium will react the fastest because it is the highest in the reactivity series.

Bullet point method (how you will do the experiment): *Include…*
- *Same volume, type, concentration and temperature of acid.*
- *Same amount of metal (mass, size of pieces).*
- *Place a mass of metal into a set volume of acid.*
- *Time the reaction, defining start and end point or measuring gas given off.*

Safety considerations:
Safe use of acid and potential heating of glass due to reaction. Lithium and acid are very reactive. Refer to COSSH.

State how you will ensure that the experiment is:
- valid – *comparing the metals, controlling other variables*
- reliable – *repeat readings*
- accurate – *how rate of reaction is measured*

EXTENSION SUGGESTION

Draw a table for your results.

WHICH METAL IS THE MOST REACTIVE?

When a metal reacts with an acid, the following reaction takes place:

metal + acid ⟶ salt + hydrogen

The more reactive a metal is, the quicker the reaction takes place.

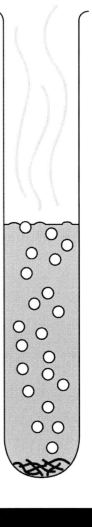

TASK

Plan an investigation to find out which of these metals – lithium, magnesium, gold and copper - is the most reactive in hydrochloric acid.

Include:
- Prediction with a scientific reason.
- Bullet point method (how you will do the experiment).
- Safety considerations.

State how you will ensure that the experiment is:
- valid
- reliable
- accurate

CHEMICAL REACTIONS
COLLECTING CARBON DIOXIDE

4

KS4 NATIONAL CURRICULUM HSW LINK

2. *Practical and enquiry skills*
   d. evaluate methods of collection of data and consider their validity and reliability as evidence

RESOURCES:
Task Sheet 4, enough for one between two learners.

TIME:
10 minutes for the activity, 10 minutes for class discussion.

NOTES

- This task is most suitable for use as a Starter, Main Activity or Plenary.
- Learners must be aware of these key words/concepts before attempting the task: evaluating data collection methods.

SUGGESTED ANSWERS

A. What are the disadvantages of each method?
   1. *Carbon dioxide can dissolve in water, so results may not be precise or accurate; temperature of water may affect volume of gas collected.*
   2. *Can't measure the volume of gas produced.*
   3. *Gas syringes are expensive.*
B. Which method would give the most accurate data?
   3. *Gas syringe method.*
C. Why can't you collect carbon dioxide using upward delivery?
   It is denser than air.

EXTENSION SUGGESTION

Describe how you would test that the gas was carbon dioxide.

CHEMICAL REACTIONS
COLLECTING CARBON DIOXIDE

4

Carbon dioxide can be made easily in the laboratory by reacting a carbonate, such as calcium carbonate, with an acid, e.g. hydrochloric acid. The carbon dioxide that is given off can be collected in a variety of ways.

Look at the diagrams of methods of collecting carbon dioxide.

1. Collecting over water into an inverted measuring cylinder:

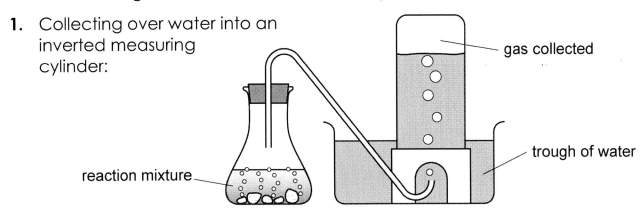

2. Collecting in an upright test tube using downward delivery:

3. Collecting in a gas syringe:

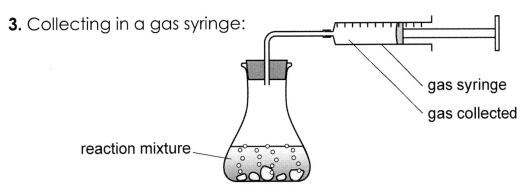

TASK

Discuss the questions below:

A. What are the disadvantages of each method?

B. Which method would give the most accurate data?

C. Why can't you collect carbon dioxide using upward delivery?

CHEMICAL REACTIONS
EVALUATING COPPER EXTRACTION

5

KS4 NATIONAL CURRICULUM HSW LINK

2. *Practical and enquiry skills*
   d. evaluate methods of collection of data and consider their validity and reliability as evidence

RESOURCES:
Task Sheet 5, enough for one between two learners.

TIME:
10-15 minutes for the activity, 10 minutes for class discussion.

NOTES

- This task is most suitable for use as a Starter, Main Activity or Plenary.
- Learners must be aware of these key words/concepts before attempting the task: evaluating data collecting methods; electrolysis.

SUGGESTED ANSWERS

For each set of results, discuss whether you think the results are accurate, precise or reliable. Decide 'Yes' or 'No' and mark it in the table.

	Group 1	Group 2	Group 3
Accurate	Yes	Yes	No
Precise	Yes	No	No
Reliable	No	No	No

Check the definition of these key words in your GCSE specification.

EXTENSION SUGGESTION

What could they do to improve the reliability, accuracy and precision of their experiment?

CHEMICAL REACTIONS
EVALUATING COPPER EXTRACTION

5

A class of Year 10s were investigating how much copper they could get from copper sulphate.

What we did:
- We set up the electrolysis as shown in the diagram.
- We weighed the negative electrode at the start of the experiment.
- Then we put the electrode in the copper sulphate solution and put a current through it for exactly 60 seconds.
- Carefully, we weighed the mass of the electrode on an electronic balance.
- Then we repeated this process ten times; we had to be careful not to lose any copper particles.
- We subtracted the mass of the electrode at the start of the experiment from the mass of the electrode after each minute and put the results in our table.

TASK

For each set of results, discuss whether you think the results are accurate, precise or reliable.

Time of electrolysis (s)	Predicted mass of Cu (g)	Group 1 mass of Cu (g)	Group 2 mass of Cu (g)	Group 3 mass of Cu (g)
0	0.00	0.00	0.0	0.0
60	0.24	0.21	0.3	0.3
120	0.48	0.42	0.5	0.5
180	0.72	0.69	0.8	0.7
240	0.96	0.91	0.9	0.9
300	1.19	1.15	1.1	0.8
360	1.43	1.40	1.2	0.9
420	1.67	1.60	1.5	1.1
480	1.91	1.82	1.8	1.4
540	2.15	2.10	2.0	1.6
600	2.39	2.30	2.2	1.9

Decide 'Yes' or 'No' and mark it in the table below.

	Group 1	Group 2	Group 3
Accurate			
Precise			
Reliable			

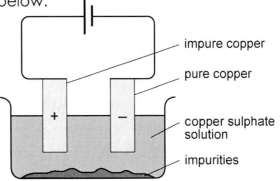

impure copper

pure copper

copper sulphate solution

impurities

KS4 NATIONAL CURRICULUM HSW LINK

3. *Communication skills*
   a. recall, analyse, interpret, apply and question scientific information or ideas
   c. present information, develop an argument and draw a conclusion, using scientific, technical and mathematical language, conventions and symbols, and ICT tools

RESOURCES:
Task Sheet 6, cut into 17 cards (1 instruction and 16 sorting cards) if used as a classroom task. Enough for one set per learner.

TIME:
15 minutes for the activity, 5 minutes for class discussion.

NOTES

This task is most suitable for use as a Starter, Main Activity or Plenary. It could also be set as a Homework Activity.

SUGGESTED ANSWERS

	Corrosive – these materials destroy living tissue.
	No naked flames – flammable or combustible chemicals may be present.
	Biohazard – living (bio) material that is harmful.
	Oxidising – these substances provide oxygen for other materials to burn.
	Highly flammable – materials that easily catch fire.
	Toxic – substances can cause death.
	Explosive – materials that can explode easily.
	Radioactive – substances that emit radiation.

EXTENSION SUGGESTION

Suggest chemicals that could be labelled with the hazard symbols.

HAZARD LABELS IN THE LABORATORY

6

TASK

Match the chemical symbol with its definition.

	Highly flammable – materials that easily catch fire.
	Toxic – substances can cause death.
	No naked flames – flammable or combustible chemicals may be present.
	Radioactive – substances that emit radiation.
	Oxidising – these substances provide oxygen for other materials to burn.
	Biohazard – living (bio) material that is harmful.
	Explosive – materials that can explode easily.
	Corrosive – these materials destroy living tissue.

CHEMICAL REACTIONS
FORENSIC TESTING

7

KS4 NATIONAL CURRICULUM HSW LINK

3. *Communication skills*
   a. how scientific data can be collected and analysed
   b. how interpretation of data, using creative thought, provides evidence to test ideas and develop theories

RESOURCES:
Task Sheet 7, enough for one between two learners.

TIME:
10 minutes for the activity, 5 minutes for class discussion.

NOTES

- This task is most suitable for use as a Starter or Plenary.
- Learners must be aware of these key words/concepts before attempting the task: data analysis; flame test; hydroxide precipitation test; precipitate.

SUGGESTED ANSWERS

A. Which method would you use first? Explain why.
You would use the flame test first because this allows you to distinguish four out of the six chemicals.

B. Which method would you use second? Explain why.
You would use the test with excess sodium hydroxide second. This would enable you to distinguish between magnesium chloride and aluminium chloride.

C. Why would the table not guarantee success with identifying the unknown chemical?
The table does not contain an exhaustive list of chlorides; flame tests have to be carried out very carefully to achieve accurate results.

EXTENSION SUGGESTION

How many tests would have to be carried out to identify all the chemicals mentioned in the table?

CHEMICAL REACTIONS
FORENSIC TESTING

A science technician found a bottle that contained a salt-like substance. However, the name of the chemical had been rubbed off the bottle and the only word that could be made out was the word 'chloride'.

The technician researched 'chlorides' on the internet and put together the following information on how chlorides react with different forensic tests.

Name of chloride	Method 1 Testing with a little sodium hydroxide	Method 2 Testing with excess sodium hydroxide	Method 3 Flame test
Calcium chloride	White precipitate forms	White precipitate forms and remains	Red
Magnesium chloride	White precipitate forms	White precipitate forms and remains	No colour
Aluminium chloride	White precipitate forms	White precipitate forms then dissolves	No colour
Zinc chloride	White precipitate forms	White precipitate forms then dissolves	White green
Potassium chloride	No precipitate	No precipitate	Lilac
Sodium chloride	No precipitate	No precipitate	Yellow

TASK

The technician decided to carry out some tests to determine the identity of the chemical, but time was limited.

Discuss these questions with a partner.

A. Which method would you use first? Explain why.

B. Which method would you use second? Explain why.

C. Why would the table not guarantee success with identifying the unknown chemical?

CHEMICAL REACTIONS
BREWING BEER

8

KS4 NATIONAL CURRICULUM HSW LINK

3. *Communication skills*
   a. recall, analyse, interpret, apply and question scientific information or ideas

RESOURCES:
Task Sheet 8, cut into 13 cards (1 instruction and 12 sorting cards) if used as a classroom task. One set per learner.

TIME:
10 minutes for the activity, 10 minutes for class discussion.

NOTES

- This task is most suitable for use as a Starter, Main Activity or Plenary. It would also be suitable as a Homework Activity.
- Learners must be aware of this key word before attempting the task: fermentation.

SUGGESTED ANSWERS

Green buds are sprouted from barley, roasted and then crushed to form a coarse powder called 'grist'.	
Hot water is mixed with the grist in a large container called a 'mash tun'.	
Sugars in the malt dissolve in the hot water to form a sweet brown liquid called 'sweet wort'.	
Hops are added to the wort and the mixture is boiled. The hops add bitterness and the wort is now called 'hopped wort'. The hops are removed.	
The cooled hopped wort is run into a fermentation vessel and yeast is added. Different types of yeast are used depending on the type of beer being brewed.	
During fermentation, the yeast converts sugar into alcohol and carbon dioxide.	
The beer goes into a cask where more hops may be added for extra taste. Finings clear the beer. A second fermentation may take place.	
Finally, the beer is pasteurised and bottled.	

EXTENSION SUGGESTION

Write out any new words and their meanings.

TASK

Beer in one form or another has been brewed for thousands of years. Today, beer is brewed industrially in stainless steel fermenters.

Read the statements and arrange them in the correct order.

Hot water is mixed with the grist in a large container called a 'mash tun'.	
Hops are added to the wort and the mixture is boiled. The hops add bitterness and the wort is now called 'hopped wort'. The hops are removed.	
Sugars in the malt dissolve in the hot water to form a sweet brown liquid called 'sweet wort'.	
During fermentation, the yeast converts sugar into alcohol and carbon dioxide.	
The beer goes into a cask where more hops may be added for extra taste. Finings clear the beer. A second fermentation may take place.	
Finally, the beer is pasteurised and bottled.	
The cooled hopped wort is run into a fermentation vessel and yeast is added. Different types of yeast are used depending on the type of beer being brewed.	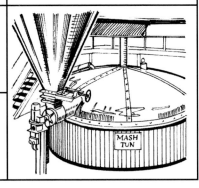
Green buds are sprouted from barley, roasted and then crushed to form a coarse powder called 'grist'.	

CHEMICAL REACTIONS
HABER: HERO OR HORRID?

9

KS4 NATIONAL CURRICULUM HSW LINK

4. *Applications and implications of science*
   c. to consider how and why decisions about science and technology are made, including those that raise ethical issues, and about the social, economic and environmental effects of such decisions

RESOURCES:
Task Sheet 9, cut into 9 cards (1 instruction and 8 sorting cards), one set between two or four learners.

TIME:
10 minutes for the activity, 10 minutes for class discussion.

NOTES

- This task is suitable for use as an extended Starter or Plenary, or a Main Activity.
- Learners must be aware of these key words/concepts before attempting the task: ethical issues; the significance of the Haber Process.

SUGGESTED ANSWERS

The statements should raise discussion about:

- Whether scientists have any responsibility for what their discoveries are used for.
- Whether science should be used to defend what you believe is right.
- Whether scientists should be judged for what their discoveries are used for.
- To what extent Haber was responsible for the use of his discoveries: fertiliser, chemical weapons, Zyklon insecticide.

EXTENSION SUGGESTION

With the evidence provided, decide whether Haber was a hero or horrid, justifying your claim.

TASK

Discuss each statement card and decide whether it is evidence for Fritz Haber, the Jewish German chemist, being a hero or a horrid person.

Fritz Haber was awarded the Nobel Prize for Chemistry in 1918 for his work on the fixation of nitrogen from the air.	Haber developed gas masks to protect soldiers from the effects chemical warfare.
In World War I, Haber developed chemical weapons to help his country.	Haber's wife committed suicide after he had released the poison gas into the trenches. She was opposed to chemical warfare.
Haber's process of nitrogen fixation allowed artificial fertilisers to be produced, so crops could be grown more easily around the world.	Haber produced the cyanide based pesticide, Zyklon B. This was later used by the Nazis to gas Jews in Germany.
Haber developed the deadly gases for trench warfare and personally oversaw their release into the enemy trenches.	Haber defended his development of chemical weapons, saying that death was death by whatever means it was inflicted.

CHEMICAL REACTIONS
METALS DISCOVERY TIMELINE

10

KS4 NATIONAL CURRICULUM HSW LINK

4. *Applications and implications of science*
   c. how uncertainties in scientific knowledge and scientific ideas change over time and about the role of the scientific community in validating these changes

RESOURCES:
Task Sheet 10, cut into 14 cards (1 instruction, 1 timeline and 12 sorting cards), one set per learner. Scissors and glue.

TIME:
20-30 minutes for the activity, 10 minutes for class discussion.

NOTES

- This task is suitable for use as a Main or Homework Activity.
- Learners must be aware of these key words/concepts before attempting the task: how scientific ideas change; properties of metals; extraction of metals.

SUGGESTED ANSWERS

The metal that was first discovered was gold; its use can be traced back to 6000 BC. The other metals that were discovered early in human history were silver, copper, mercury, tin, iron and lead. These were discovered mostly in their natural form, apart from iron, which was smelted from its ore using charcoal.
These metals were the only metals known until the 1400s AD, when arsenic was discovered by Albertus Magnus. The Chinese were using zinc during this time.
Platinum was discovered in the 1500s in Mexico. In 1595, bismuth was discovered by the reduction of its ore using carbon, though it was not until 1753 that bismuth was classified as an element.
In the 1700s, cobalt, nickel, manganese, molybdenum, tungsten, uranium, zirconium and yttrium were extracted using carbon, and tungsten was the first metal to be extracted using hydrogen.
Until the 1800s, metals could only be extracted using carbon or hydrogen. It was not until scientists started to dabble with electricity that more reactive metals could be separated from their ores. Sir Humphrey Davy was the first to separate potassium and sodium using electrodes.
It was not until electrolysis was used on a large scale that aluminium was able to be extracted and used. It is belived that Napoleon gave a banquet where the most honoured guests were given aluminium utensils, while the other guests had to make do with gold ones.
Dmitri Mendeleev published the first Periodic Table in 1869, containing the 60 known elements at that time. He left gaps for undiscovered elements and the discovery of the metals gallium in 1875, scandium in 1879 and germanium in 1886 supported his predictions.
In the 1900s, a further 20 or so elements were discovered. These metals were found in the Earth's crust in very small amounts and some may be so unstable, they exist for only a few seconds when extracted. These included neptunium, plutonium, curium, einsteinium, fermium, mendelevium and nobelium.

EXTENSION SUGGESTION

Add your own stages. For example, who discovered uranium and when? What is the future of metal discovery?

METALS DISCOVERY TIMELINE

10

TASK

Cut out the timeline, the statements and the images. Read each statement carefully and place it on the timeline. Match the images to the timeline to illustrate it.

Timeline

— 0 AD

— 1200 AD

— 1400 AD

— 1600 AD

— 1800 AD

— 2000 AD

Until the 1800s, metals could only be extracted using carbon or hydrogen. It was not until scientists started to dabble with electricity that more reactive metals could be separated from their ores. Sir Humphrey Davy was the first to separate potassium and sodium using electrodes.

These metals were the only metals known until the 1400s AD, when arsenic was discovered by Albertus Magnus. The Chinese were using zinc during this time.

The metal that was first discovered was gold; its use can be traced back to 6000 BC. The other metals that were discovered early in human history were silver, copper, mercury, tin, iron and lead. These were discovered mostly in their natural form, apart from iron, which was smelted from its ore using charcoal.

Dmitri Mendeleev published the first Periodic Table in 1869, containing the 60 known elements at that time. He left gaps for undiscovered elements and the discovery of the metals gallium in 1875, scandium in 1879 and germanium in 1886 supported his predictions.

It was not until electrolysis was used on a large scale that aluminium was able to be extracted and used. It is belived that Napoleon gave a banquet where the most honoured guests were given aluminium utensils, while the other guests had to make do with gold ones.

In the 1900s, a further 20 or so elements were discovered. These metals were found in the Earth's crust in very small amounts and some may be so unstable, they exist for only a few seconds when extracted. These included neptunium, plutonium, curium, einsteinium, fermium, mendelevium and nobelium.

Platinum was discovered in the 1500s in Mexico. In 1595, bismuth was discovered by the reduction of its ore using carbon, though it was not until 1753 that bismuth was classified as an element.

In the 1700s, cobalt, nickel, manganese, molybdenum, tungsten, uranium, zirconium and yttrium were extracted using carbon, and tungsten was the first metal to be extracted using hydrogen.

carbon anodes
solution of aluminium oxide in molten cryolite
steel tank lined with refractory bricks
molten aluminium collects at the bottom
carbon as cathode

CHEMICAL PATTERNS
MENDELEEV'S PREDICTIONS

11

KS4 NATIONAL CURRICULUM HSW LINK

1. *Data, evidence, theories and explanations*
   b. how interpretation of data, using creative thought, provides evidence to test ideas and develop theories
   c. how explanations of many phenomena can be developed using scientific theories, models and ideas

RESOURCES:
Task Sheet 11, enough for one per learner.

TIME:
10 minutes for the activity, 5 minutes for class discussion.

NOTES

- This task is most suitable for use as a Starter or Plenary.
- Some literacy support may be required.
- Learners must be aware of these key words/concepts before attempting the task: making predictions; the Periodic Table; density; atomic mass.

SUGGESTED ANSWERS

Decide which element 'eka-silicon' is most likely to be. Give reasons.
Germanium – closest in properties.
(Gallium – melting point too low and density of chloride too high.)
(Arsenic discovered in 1250! Also, chloride salt melting point too low.)

EXTENSION SUGGESTION

What is the difference between atomic mass and atomic number?

The Periodic Table is the most valuable tool a chemist has to hand. It lists all the known elements in order of atomic number, while keeping elements with similar properties grouped together.

The Periodic Table that we use today was conceived by a Russian chemist called Dmitri Mendeleev (1822-1907). He made a card for each known element, listing all its properties. He then arranged the elements by their atomic mass and properties. What made Mendeleev's table different from other efforts to order the elements was that he left gaps in his table where elements were missing. He was able to accurately predict the properties of these missing elements. One such element he made predictions for was 'eka-silicon'.

Mendeleev predicted that 'eka-silicon' would have an atomic mass of 72, with a density of 5.5 g/cm³ and a high melting point. He thought the colour of this element would be grey. It would form a chloride salt with a density of 1.9 g/cm³ and a boiling point of less than 100°C.

TASK

Read the text above. Using the information in the text, decide which element 'eka-silicon' is most likely to be. Give reasons for your answer.

Element name	Discovered	Atomic mass	Melting point (°C)	Density (g/cm³)	Chloride salt density	Chloride salt boiling point (°C)	Colour
Gallium	1875	70	30	5.9	2.5	78	Grey
Germanium	1886	73	947	5.3	1.9	86	Grey
Arsenic	1250	75	613	5.7	2.2	−16	Grey

CHEMICAL PATTERNS
ELEMENT SORT

12

KS4 NATIONAL CURRICULUM HSW LINK

1. *Data, evidence, theories and explanations*
   a. how scientific data can be collected and analysed
   b. how interpretation of data, using creative thought, provides evidence to test ideas and develop theories

RESOURCES:
Task Sheet 12, cut into 13 cards (1 instruction and 12 sorting cards), enough for one set between two to four learners. Periodic Tables.

TIME:
15 minutes for the activity, 5 minutes for class discussion.

NOTES

- This task is most suitable for use as a Starter, Main Activity or Plenary.
- Learners must be aware of these key words/concepts before attempting the task: the Periodic Table arrangement based on atomic number and electron shells.

SUGGESTED ANSWERS

Element 1	Lithium (Group 1)
Element 2	Iron (Transition metals)
Element 3	Bromine (Group 7)
Element 4	Sodium (Group 1)
Element 5	Mercury (Transition metals)
Element 6	Helium (Group 8/0)
Element 7	Potassium (Group 1)
Element 8	Fluorine (Group 7)
Element 9	Neon (Group 8/0)
Element 10	Copper (Transition metals)
Element 11	Chlorine (Group 7)
Element 12	Argon (Group 8/0)

EXTENSION SUGGESTION

Write the chemical symbol for each element.

TASK

Read the information about each element carefully.

Step 1 Sort the cards into four groups: Group 1 elements, Group 7 elements, Group 8/0 elements and transition elements.

Step 2 Decide the name of each element, using the Periodic Table to help if you need to.

Step 3 Arrange the cards as they would be in the Periodic Table.

ELEMENT 1

Metal
Solid
(at room temperature)
I have 3 electrons.

ELEMENT 2

Metal
Solid
(at room temperature)
I am a dense metal that can sometimes be magnetic.

ELEMENT 3

Non-metal
Liquid
(at room temperature)
I am so unusual, I am not going to give you a clue!

ELEMENT 4

Metal
Solid
(at room temperature)
I react strongly with Element 11, making a salt.

ELEMENT 5

Metal
Liquid
(at room temperature)
I am toxic.

ELEMENT 6

Non-metal
Gas
(at room temperature)
I am lighter than air and very hard to react with.

ELEMENT 7

Metal
Solid
(at room temperature)
I react violently with water by floating on the surface and burning in a violet flame.

ELEMENT 8

Non-metal
Gas
(at room temperature)
My outer electron shell has one electron missing, so I am very reactive.

ELEMENT 9

Non-metal
Gas
(at room temperature)
My electron arrangement is a perfect 2, 8, 8.

ELEMENT 10

Metal
Solid
(at room temperature)
My oxide is black and my sulphate is bright blue.

ELEMENT 11

Non-metal
Gas
(at room temperature)
My outer shell has one electron missing, so I am quite reactive.

ELEMENT 12

Non-metal
Gas
(at room temperature)
I am so unreactive that I can stop light bulb filaments from burning.

MAKING MOLAR SOLUTIONS

13

KS4 NATIONAL CURRICULUM HSW LINK

2. *Practical and enquiry skills*
   c. work accurately and safely, individually and with others, when collecting first-hand data
   d. evaluate methods of collection of data and consider their validity and reliability as evidence

RESOURCES:
Task Sheet 13, cut into 9 cards (1 instruction and 8 sorting cards), enough for one set between two or four learners.

TIME:
10 minutes for the activity, 5 minutes for class discussion.

NOTES

- This task is most suitable for use as a Starter, Main Activity or Plenary. It is also suitable as a Homework Activity.
- Learners must be aware of these key words/concepts before attempting the task: molar solutions; atomic mass; calculating molecular mass.

SUGGESTED ANSWERS

0.1M Solutions B & H
0.5M Solutions F & G
1.0M Solutions A & C
2.0M Solutions D & E

EXTENSION SUGGESTION

Consider the practical problems of preparing each of these solutions.

TASK

A molar solution is an exact number of atoms (one mole) dissolved in one litre of water. A mole is equivalent to the molecular mass of the molecule.

Read each card and decide if it is describing a 0.1M, 0.5M, 1.0M or 2.0M solution of either sodium chloride or sodium hydroxide.

Atomic mass: H = 1, O = 16, Na = 23, Cl = 35.5

Solution A	Solution E
58.5g of sodium chloride (NaCl) is dissolved in 1 litre of water.	117g of sodium chloride (NaCl) is dissolved in 1 litre of water.
Solution B	**Solution F**
5.85g of sodium chloride (NaCl) is dissolved in 1 litre of water.	29.25g of sodium chloride (NaCl) is dissolved in 1 litre of water.
Solution C	**Solution G**
40g of sodium hydroxide (NaOH) is dissolved in 1 litre of water.	40g of sodium hydroxide (NaOH) is dissolved in 2 litres of water.
Solution D	**Solution H**
40g of sodium hydroxide (NaOH) is dissolved in 500cm³ of water	40g of sodium hydroxide (NaOH) is dissolved in 10 litres of water.

CHEMICAL PATTERNS
THE SIZE OF ATOMS

14

KS4 National Curriculum HSW link

RESOURCES:
Task Sheet 14, enough for one per learner. Graph paper.

TIME:
15 minutes for the activity, 5 minutes for class discussion.

3. *Communication skills*
   a. recall, analyse, interpret, apply and question scientific information or ideas
   b. use both qualitative and quantitative approaches
   c. present information, develop an argument and draw a conclusion, using scientific, technical and mathematical language, conventions and symbols, and ICT tools

NOTES

- This task is most suitable for use as a Starter or Plenary.
- Learners must be aware of these key words/concepts before attempting the task: Periodic Table; Group 1 metals; using line graphs to make predictions.

SUGGESTED ANSWERS

A. Decide on the independent and dependent variables.
Independent variable: atomic number. Dependent variable: atomic radius.

B/C. Graph should look something like this:

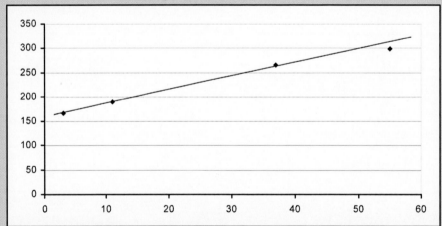

D. Using your graph, estimate the atomic radius for potassium.
243 is the value, but allow 233-246 from a good line of best fit.

EXTENSION SUGGESTION

How confident are you in your estimate?

The radius of atoms can be calculated.

In Table 1 below, the atomic radii of the elements in Group 1 of the Periodic Table are listed. However, the atomic radius for potassium is missing.

Use your graph drawing skills to estimate the atomic radius of potassium.

Table 1: The atomic radii of the Group 1 elements

Element	Atomic number	Atomic radius (picometres)
Lithium	3	167
Sodium	11	190
Potassium	19	?
Rubidium	37	265
Caesium	55	298

TASK

A. Decide on the independent and dependent variables.

B. Plot a suitable line graph for this data.

C. Draw a line of best fit.

D. Using your graph, estimate the atomic radius for potassium.

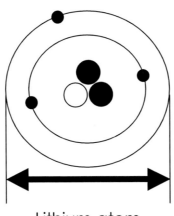

Lithium atom
radius 167 pm

CHEMICAL PATTERNS
ELEMENT MELTING POINT PATTERNS

15

KS4 NATIONAL CURRICULUM HSW LINK

3. *Communication skills*
   a. recall, analyse, interpret, apply and question scientific information or ideas
   c. present information, develop an argument and draw a conclusion, using scientific, technical and mathematical language, conventions and symbols, and ICT tools

RESOURCES:
Task Sheet 15, enough for one per learner, graph paper.

TIME:
15 minutes for the activity, 5 minutes for class discussion.

NOTES

- This task is most suitable for use as a Main or Homework Activity.
- Learners must be aware of these key words/concepts before attempting the task: graph drawing; the Periodic Table; atomic number; melting point.

SUGGESTED ANSWERS

A. Decide how to best present the data.
   Learners should discuss the pros and cons of a bar chart or line graph.

B. Present the data, deciding on the independent and dependent variables.
   Independent variable = atomic number. Dependent variable = melting point.

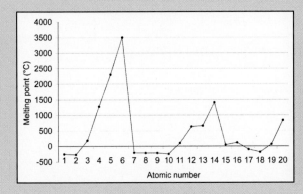 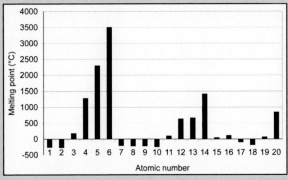

EXTENSION SUGGESTION

Describe the pattern in the graph.

ELEMENT MELTING POINT PATTERNS

15

The table below shows the melting points of the first twenty elements in the Periodic Table:

Element name	Atomic number	Melting point °C
Hydrogen	1	−259
Helium	2	−272
Lithium	3	180
Beryllium	4	1278
Boron	5	2300
Carbon	6	3500
Nitrogen	7	−210
Oxygen	8	−218
Fluorine	9	−219
Neon	10	−248
Sodium	11	98
Magnesium	12	630
Aluminium	13	660
Silicon	14	1410
Phosphorous	15	44
Sulphur	16	113
Chlorine	17	−101
Argon	18	−189
Potassium	19	64
Calcium	20	839

TASK

A. Decide how to best present the data.

B. Present the data, deciding on the independent and dependent variables.

CARBON CHAINS AND BOILING POINTS

16

KS4 NATIONAL CURRICULUM HSW LINK

3. *Communication skills*
   a. recall, analyse, interpret, apply and question scientific information or ideas
   b. use both qualitative and quantitative approaches
   c. present information, develop an argument and draw a conclusion, using scientific, technical and mathematical language, conventions and symbols, and ICT tools

RESOURCES:
Task Sheet 16, enough for one per learner, graph paper.

TIME:
15 minutes for the activity, 5 minutes for class discussion.

NOTES

- This task is most suitable for use as a Main or Homework Activity.
- Learners must be aware of these key words/concepts before attempting the task: graph drawing; using data to make predictions.

SUGGESTED ANSWERS

A/B. Learners should draw a line graph and line of best fit.

C. Use the graph to predict the boiling point of dodecane – an alkane of 12 carbons.
   Should be about 216°C.

D. Use the graph to predict the boiling point of diesel fuel – a mixture of alkanes of between 14-20 carbons.
   About 290-340°C.

E. Use the graph to predict the length of the carbon chain of an alkane with a boiling point of 400°C.
   Refer to their graph, but should be about 25 carbons.

EXTENSION SUGGESTION

Suggest reasons why boiling point increases with chain length.

CHEMICAL PATTERNS
CARBON CHAINS AND BOILING POINTS

16

Scientific data can be used to make predictions. Use this data to make predictions about the boiling points of hydrocarbons.

Length of carbon chain	Boiling point (°C)
5	36
6	69
7	98
8	125
9	151
10	174
20	343
30	450

TASK

A. Draw a suitable graph for this data.

B. Draw a line of best fit.

C. Use the graph to predict the boiling point of dodecane – an alkane of 12 carbons.

D. Use the graph to predict the boiling point of diesel fuel – a mixture of alkanes of between 14-20 carbons.

E. Use the graph to predict the length of the carbon chain of an alkane with a boiling point of 400°C.

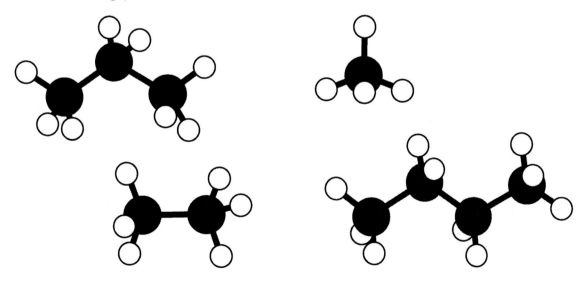

CHEMICAL PATTERNS
MATCHING MOLECULES

17

KS4 National Curriculum HSW link

3. *Communication skills*

   c. present information, develop an argument and draw a conclusion, using scientific, technical and mathematical language, conventions and symbols, and ICT tools

RESOURCES:
Task Sheet 17, cut into 28 cards (1 instruction and 27 sorting cards), enough for one set per learner.

TIME:
10 minutes for the activity, 5 minutes for class discussion.

Notes

- This task is most suitable for use as a Starter or Plenary.
- Learners must be aware of these key words/concepts before attempting the task: chemical formulae of common molecules.

Suggested answers

CO	Carbon monoxide	
CO_2	Carbon dioxide	
CH_4	Methane	
HCl	Hydrochloric acid	
NH_3	Ammonia	
H_2O	Water	
H_2SO_4	Sulphuric acid	
$NaOH$	Sodium hydroxide	
$NaCl$	Sodium chloride	

Extension suggestion

Put the chemicals in order of fewest to most atoms.

TASK

Match the chemical formula with the chemical name and correct diagram.

	CO	Sulphuric acid
	CO_2	Sodium hydroxide
	CH_4	Sodium chloride
	HCl	Water
	NH_3	Carbon monoxide
	H_2O	Carbon dioxide
	H_2SO_4	Methane
	NaOH	Hydrochloric acid
	NaCl	Ammonia

ENERGY CHANGING CHEMICAL REACTIONS

18

RESOURCES:
Task Sheet 18, enough for one per learner.

TIME:
10 minutes for the activity, 5 minutes for class discussion.

KS4 NATIONAL CURRICULUM HSW LINK

3. *Communication skills*
   a. recall, analyse, interpret, apply and question scientific information or ideas
   b. use both qualitative and quantitative approaches
   c. present information, develop an argument and draw a conclusion, using scientific, technical and mathematical language, conventions and symbols, and ICT tools

NOTES

- This task is most suitable for use as a Starter, Main Activity or Plenary. It is also suitable as a Homework Activity.
- Learners must be aware of these key words/concepts before attempting the task: exothermic; endothermic; collecting reliable data; fair testing.

SUGGESTED ANSWERS

A. Calculate the temperature differences.
   Sodium chloride: no change
   Calcium chloride: +4°C
   Ammonium chloride: −6°C

B. Decide whether the reactions are exothermic or endothermic.
   Sodium chloride: neither
   Calcium chloride: exothermic
   Ammonium chloride: endothermic

C. Identify the independent and dependent variables.
   Independent variable: type of salt
   Dependent variable: temperature change

D. Suggest how they could make their investigation more reliable.
   Repeat the experiment.

E. Discuss whether their investigation was a fair test and how it could be improved.
   Same volume of water, same time before final temperature was taken.
   Improve by weighing the salts instead of using a spatula.

EXTENSION SUGGESTION

Suggest reasons why a data logger would be useful in this type of investigation.

Mac and Teagan carried out an investigation to see whether there was a temperature change when they added water to three different salts.

They put a thermometer in a test tube and added one spatula of the salt. Then they added 5cm³ of water and read the thermometer. This was recorded as the start temperature. They stirred the water and salt with the thermometer and recorded the temperature again after a minute. They put their results in the table below.

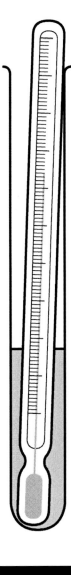

Salt	Temperature (°C)		
	Start	**Finish**	**Difference**
Sodium chloride	18	18	
Calcium chloride	18	22	
Ammonium chloride	18	12	

TASK

A. Calculate the temperature differences.

B. Decide whether the reactions are exothermic or endothermic.

C. Identify the independent and dependent variables.

D. Suggest how they could make their investigation more reliable.

E. Discuss whether their investigation was a fair test and how it could be improved.

MOBILE METALS: BENEFITS, DRAWBACKS AND RISKS

19

KS4 NATIONAL CURRICULUM HSW LINK

4. *Applications and implications of science*
   a. about the use of contemporary scientific and technological developments and their benefits, drawbacks and risks

RESOURCES:
Task Sheet 19, cut into 9 cards (1 instruction and 8 sorting cards), enough for one set per learner.

TIME:
10 minutes for the activity, 10 minutes for class discussion.

NOTES

- This task is suitable for use as an extended Starter or Plenary, or a Main Activity.
- Learners must be aware of these key words/concepts before attempting the task: benefit; drawback; risk.

SUGGESTED ANSWERS

Below is the intention of each statement: Benefit (B), Drawback (D) or Risk (R).
- Tantulum is very useful in making mobile phones. It has a very high melting point, is an excellent conductor and is resistant to corrosion. *(B)*
- The main source of the metal tantulum is a **rare** mineral called columbite tantalite, also known as coltan. *(D)*
- The richest reserves of coltan are found in Africa. 80% is found in the **forests** of the Democratic Republic of Congo. *(D)*
- Miners of coltan are eating bush meat, including endangered species such as the Drill, a primate that could become extinct within 10 years. *(D)*
- Farmers in the Congo have turned to mining for coltan in the forests, making much more money from selling the mineral. *(B)*
- Mining for coltan, the mineral that contains tantalum, is causing deforestation. *(D)*
- The effects of tantalum on human health are unknown, some small scale studies suggest it may cause tumours in animals. *(R)*
- Tantulum powder is used in capacitors and high-power resistors. Its discovery has led to smaller and lighter electrical components, ideal for mobile phones. *(B)*

EXTENSION SUGGESTION

List any other benefits, drawbacks or risks you can think of.

TASK

Discuss each statement card and decide whether it is a benefit, drawback or a risk of the use of the metal tantalum in mobile phones.

- A **benefit** is something that generally has a good effect on people.
- A **drawback** is something that is a hindrance or is the 'downside'.
- A **risk** is a possible danger or source of harm.

Tantulum is very useful in making mobile phones. It has a very high melting point, is an excellent conductor and is resistant to corrosion.	The main source of the metal tantalum is a rare mineral called *col*umbite *tan*talite, also known as **coltan**.
The richest reserves of coltan are found in Africa. 80% is found in the forests of the Democratic Republic of Congo.	Miners of coltan are eating bush meat, including endangered species such as the Drill, a primate that could become extinct within 10 years.
Farmers in the Congo have turned to mining for coltan in the forests, making much more money from selling the mineral.	Mining for coltan, the mineral that contains tantalum, is causing deforestation.
The effects of tantalum on human health are unknown, some small scale studies suggest it may cause tumours in animals.	Tantulum powder is used in capacitors and high-power resistors. Its discovery has led to smaller and lighter electrical components, ideal for mobile phones.

STRUCTURE OF THE ATOM TIMELINE

KS4 NATIONAL CURRICULUM HSW LINK

4. *Applications and implications of science*
   c. how uncertainties in scientific knowledge and scientific ideas change over time and about the role of the scientific community in validating these changes

RESOURCES:
Task Sheet 20, cut into 14 cards (1 instruction, 1 timeline and 12 sorting cards), enough for one set per learner.

TIME:
20-30 minutes for the activity, 10 minutes for class discussion.

NOTES

- This task is suitable for use as a Main or Homework Activity.
- Learners must be aware of these key words/concepts before attempting the task: how scientific ideas change.

SUGGESTED ANSWERS

Around 442 BC, Democritus, a Greek philosopher, proposes an atomic theory of the universe, saying that all matter is made up of atoms which are eternal, invisible and so small they can't be divided.	
J. Dalton (1803) develops an 'atomic theory' in which atoms are tiny, indestructible, solid spheres.	
In 1897, J.J. Thomson proposes the 'plum pudding' model to explain the structure of the atom. He also discovers the electron.	
Nagoaka (1903) proposes a model in which the atom resembles the planet Saturn, i.e. the electrons orbit a central nucleus in a single plane.	
In 1911, E. Rutherford states that the mass of the atom is in a small positively charged ball at the centre, surrounded by a cloud of electrons. Rutherford is also credited with discovering protons.	Nucleus
H.G.J. Moseley says that the 'atomic number' of an element equals the number of protons in the nucleus. The Periodic Table is laid out in order of number rather than atomic mass (1914).	Electron cloud
N. Bohr (1922) proposes that electrons are arranged in successively large orbits around the nucleus. This theory explains the regularities seen in the Periodic Table.	
In 1932, Chadwick discovers the neutron, a neutrally charged particle with a mass similar to that of a proton.	

EXTENSION SUGGESTION

Suggest why each of these discoveries happened in this order.

STRUCTURE OF THE ATOM TIMELINE

20

TASK

Cut out the timeline, the statements and the images. Read each statement carefully and place it on the timeline. Match the images to the timeline to illustrate it.

Timeline		
	Nagoaka (1903) proposes a model in which the atom resembles the planet Saturn, i.e. the electrons orbit a central nucleus in a single plane.	
	J. Dalton (1803) develops an 'atomic theory' in which atoms are tiny, indestructible, solid spheres.	
— 1700 AD		
	In 1911, E. Rutherford states that the mass of the atom is in a small positively charged ball at the centre, surrounded by a cloud of electrons. Rutherford is also credited with discovering protons.	
— 1800 AD	Around 442 BC, Democritus, a Greek philosopher, proposes an atomic theory of the universe, saying that all matter is made up of atoms which are eternal, invisible and so small they can't be divided.	
	N. Bohr (1922) proposes that electrons are arranged in successively large orbits around the nucleus. This theory explains the regularities seen in the Periodic Table.	
— 1900 AD	In 1897, J.J. Thomson proposes the 'plum pudding' model to explain the structure of the atom. He also discovers the electron.	
— 2000 AD	In 1932, Chadwick discovers the neutron, a neutrally charged particle with a mass similar to that of a proton.	
	H.G.J. Moseley says that the 'atomic number' of an element equals the number of protons in the nucleus. The Periodic Table is laid out in order of number rather than atomic mass (1914).	Nucleus Electron cloud

KS4 NATIONAL CURRICULUM HSW LINK

1. *Data, evidence, theories and explanations*
   a. how scientific data can be collected and analysed
   b. how interpretation of data, using creative thought, provides evidence to test ideas and develop theories

RESOURCES:
Task Sheet 21, enough for one per learner.

TIME:
10 minutes for the activity, 10 minutes for class discussion.

NOTES

- This task is most suitable for use as an extended Starter, or a Main or Homework Activity.
- Learners must be aware of these key words/concepts before attempting the task: recycling; graph interpretation; independent and dependent variables.

SUGGESTED ANSWERS

A. The percentage energy saved by recycling glass is greater than for recycling aluminium.
   Not supported.

B. Recycling aluminium, paper and glass saves energy.
   Supported.

C. Glass requires more energy than paper to make it from its raw materials.
   Cannot tell.

D. Recycling aluminium should be a greater priority for reducing global warming than recycling paper.
   Supported (although hopefully will stimulate discussion about deforestation, use of managed forests, etc).

EXTENSION SUGGESTION

Suggest what additional evidence you would require to answer the 'cannot tell' statement.

Figure 1 below shows the amount of energy saved by recycling three resources compared to obtaining the resource from its raw materials.

For example, to get aluminium from its raw material, bauxite has to be mined and transported to an aluminium extraction site. Aluminium can only be extracted from bauxite by using electrolysis, which uses a lot of energy. Recycling aluminium cans just requires collection, cleaning, melting and reforming, a process that requires 95% less energy than extracting aluminium from bauxite.

Figure 1: **Comparing the percentage of energy saved by recycling resources compared to obtaining the resources from their raw material.**

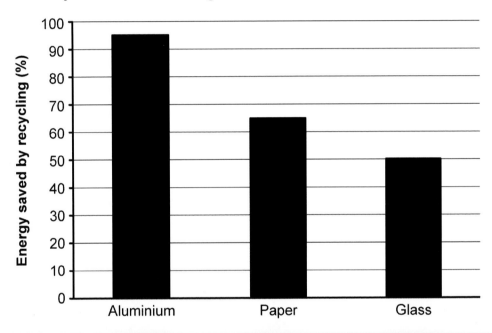

TASK

Using **only** the graph and the information above, discuss and decide whether the following statements are **supported**, **not supported** by the information, or you **cannot tell**.

A. The percentage energy saved by recycling glass is greater than for recycling aluminium.

B. Recycling aluminium, paper and glass saves energy.

C. Glass requires more energy than paper to make it from its raw materials.

D. Recycling aluminium should be a greater priority for reducing global warming than recycling paper.

NATURAL RESOURCES
MAKING BIOFUELS

22

KS4 NATIONAL CURRICULUM HSW LINK

2. *Practical and enquiry skills*
   a. plan to test a scientific idea, answer a scientific question or solve a scientific problem

RESOURCES:
Task Sheet 22, cut into 9 cards (1 instruction and 8 sorting cards), enough for one set per learner or one between two.

TIME:
10 minutes for the activity, 5 minutes for class discussion.

NOTES

- This activity is most suitable for use as a Starter, Main Activity or Plenary.
- Learners must be aware of these key words/concepts before attempting the task: lye; methanol; biofuel.

SUGGESTED ANSWERS

Mix lye (potassium hydroxide or sodium hydroxide) with methanol.
Swirl the lye with the methanol until the hydroxide dissolves and methoxide forms.
Heat the waste (vegetable) oil to 55ºC.
Combine the heated oil with the methoxide. Maintaining a constant temperature of 55ºC, stir the mixture for one hour.
Decant the oil/methoxide mixture and allow it to settle for 12-24 hours.
Glycerine, a waste product, will form a layer at the bottom of the container. The biofuel forms above the glycerine.
Decant the biofuel into a clean container and test it for quality.
Wash the biofuel with water. When the biofuel has cleared, it is ready to use.

EXTENSION SUGGESTION

What safety precautions would you have to take when making biofuel at home?

MAKING BIOFUELS

22

TASK

Rearrange these cards into the correct order to show how a biofuel can be processed from vegetable oil.

Swirl the lye with the methanol until the hydroxide dissolves and methoxide forms.

Mix lye (potassium hydroxide or sodium hydroxide) with methanol.

Wash the biofuel with water. When the biofuel has cleared, it is ready to use.

Combine the heated oil with the methoxide. Maintaining a constant temperature of 55°C, stir the mixture for one hour.

Decant the biofuel into a clean container and test it for quality.

Glycerine, a waste product, will form a layer at the bottom of the container. The biofuel forms above the glycerine.

Heat the waste (vegetable) oil to 55°C.

Decant the oil/methoxide mixture and allow it to settle for 12-24 hours.

NATURAL RESOURCES
FUEL ECONOMY

KS4 NATIONAL CURRICULUM HSW LINK

2. *Practical and enquiry skills*
   b. collect data from primary or secondary sources, including using ICT sources and tools

RESOURCES:
Task Sheet 23, enough for one per learner or one between two.

TIME:
10 minutes for the activity, 5 minutes for class discussion.

NOTES

- This activity is most suitable for use as a Starter, Main Activity or Plenary.
- Teachers may need to read through the text with less able readers before they attempt the task.
- Learners must be aware of these key words/concepts before attempting the task: fuel economy.

SUGGESTED ANSWERS

The table learners should draw is similar to this:

Fuel	Energy economy (miles per gallon)
Petrol	35.0
Diesel	45.5
Hybrid	65.7
Petrol/Bioethanol	46.0
LPG	58.0

EXTENSION SUGGESTION

Suggest reasons why 'green cars' are not very common on our roads despite their excellent fuel economy.

FUEL ECONOMY

Fuel economy is an important consideration when buying a car. Diesel and petrol power most of the cars on our roads. Petrol cars are the most common type of car, with a fuel economy of about 35 mpg (miles per gallon). With increasing costs of fuel, many drivers are now turning to diesel engines. Diesel engines can be up to 30% more efficient compared to the equivalent petrol powered engine.

In recent years, 'green cars' have become more popular. Most 'green cars' are vehicles powered by petrol alternated with a greener fuel, e.g. LPG, bioethanol and electric motors.

The hybrid car is powered by a petrol engine combined with an electric motor. The 'combined' fuel economy of a hybrid car is around 65.7 mpg. This is about 30% more economical than FFVs (flexi-fuel vehicles), which run on a mixture of petrol (15%) and bioethanol (85%).

But drivers don't have to buy a new car to be greener. Normal petrol powered vehicles can be converted to run on petrol and LPG (liquefied petroleum gas). An LPG conversion can have a fuel economy of 58 mpg.

TASK

Read the text above and put the information into a table that shows the energy efficiency of different fuels.

RESEARCHING LIMESTONE

KS4 NATIONAL CURRICULUM HSW LINK

2. *Practical and enquiry skills*
   a. plan to test a scientific idea, answer a scientific question or solve a scientific problem
   b. collect data from primary or secondary sources, including using ICT sources and tools
   c. work accurately and safely, individually and with others, when collecting first-hand data
   d. evaluate methods of collection of data and consider their validity and reliability as evidence

> **RESOURCES:**
> Task Sheet 24, enough for one per learner.
>
> **TIME:**
> 15 minutes for the activity, 10 minutes for class discussion.

NOTES

- This task is suitable for use as a Main or Homework Activity.
- Learners with reading difficulties will need support either by a peer or by the teacher reading through the text.
- Learners must be aware of these key words/concepts before attempting the task: limestone; researching secondary resources; evaluating bias.

SUGGESTED ANSWERS

The focus of this task is on the process of planning the activity. The planned activity does not need to be carried out.

What are the questions you are trying to answer?

Hopefully, the two suggested questions, where is limestone extracted and what is it used for? Also, how much limestone is used for each purpose?

What scientific evidence will you look for?

Quantitative. What is limestone used for, how much is used for each purpose?

Will you use a search engine? Which key words could you try?

Name of search engine, with reason. Key words could be anything relevant, e.g. limestone quarrying, uses of limestone UK, where limestone is quarried, etc.

Will you use the library? Which resources will you use?

National Statistics references, reference books.

How will you ensure your data is unbiased?

Know the source of the data, e.g. environmental group publications may exaggerate claims, quarrying companies may exaggerate their claims, national statistics should be unbiased, etc.

EXTENSION SUGGESTION

Carry out a search.

RESEARCHING LIMESTONE

24

- Where is limestone extracted from in the United Kingdom?
- What is the extracted limestone used for?
- How could you find out using secondary resources?

TASK

Plan to use secondary resources to find out where limestone is extracted in the UK and what it is used for.

Write down the steps you intend to take to gather scientific, reliable and accurate data.

Include:
- What are the questions you are trying to answer?
- What scientific evidence will you look for?
- Will you use a search engine? Which key words could you try?
- Will you use the library? Which resources will you use?
- How will you ensure your data is unbiased?

KS4 NATIONAL CURRICULUM HSW LINK

3. *Communication skills*
   a. recall, analyse, interpret, apply and question scientific ideas
   b. use both qualitative and quantitative approaches
   c. present information, develop an argument and draw a conclusion, using scientific, technical and mathematical language, conventions and symbols, and ICT tools

RESOURCES:
Task Sheet 25, enough for one per learner. Calculators, rulers, protractors. The task could also be carried out using a spreadsheet software program.

TIME:
20 minutes for the activity, 5 minutes for class discussion.

NOTES

- This activity is most suitable for use as an extended Starter, or a Main or Homework Activity.
- Learners must be aware of these key words/concepts before attempting the task: drawing graphs and pie charts; graph analysis.

SUGGESTED ANSWERS

A. Use the information in the table to construct a pie chart and a bar graph of the data. The pie chart will need a key.

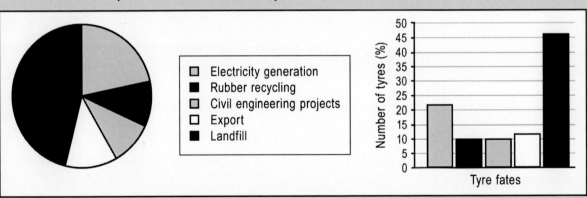

B. Which graphic best represents the data?
   The pie chart has the best representation of the data.

EXTENSION SUGGESTION

Research how tyres are recycled.

NATURAL RESOURCES
WASTE TYRE GRAPHS

Approximately 40 million tyres are removed from vehicles every year in the UK.

The tyres have various fates:

Electricity generation and other fuel resources	22%
Rubber recycling	10%
Civil engineering projects	10%
Exported	12%
Landfill	46%

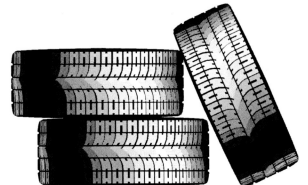

TASK

A. Use the information above to construct a pie chart and a bar graph of the data. The pie chart will need a key.

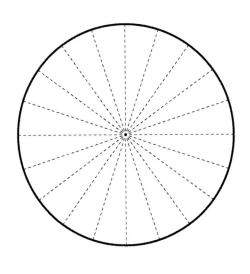

1. Pie chart

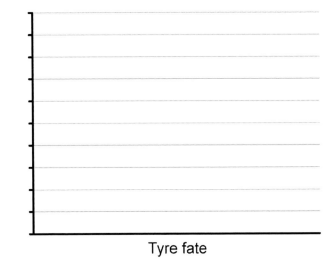

Number of tyres (%)

Tyre fate

2. Bar chart

B. Which graphic best represents the data?

NATURAL RESOURCES
FRACTIONAL DISTILLATION OF CRUDE OIL

26

KS4 NATIONAL CURRICULUM HSW LINK

3. *Communication skills*
   a. recall, analyse, interpret, apply and question scientific information or ideas
   c. present information, develop an argument and draw a conclusion, using scientific, technical and mathematical language, conventions and symbols, and ICT tools

RESOURCES:
Task Sheet 26, enough for one per learner. Graph paper.

TIME:
20 minutes for the activity, 5 minutes for class discussion.

NOTES

- This task is most suitable for use as a Main or Homework Activity, or an extended Plenary.
- Learners must be aware of these key words/concepts before attempting the task: fractional distillation; graph drawing.

SUGGESTED ANSWERS

What type of graph will you draw? *Bar graph.*
Independent variable: *chain length or fraction name.*
Dependent variable: *temperature.*

EXTENSION SUGGESTION

Describe the overall trend of the graph.

FRACTIONAL DISTILLATION OF CRUDE OIL

26

Fractional distillation is the method used to separate crude oil into fractions with different boiling points. The different fractions, e.g. petrol, diesel, have different numbers of carbon atoms linked in a chain.

TASK

Draw a graph to represent the data shown in the diagram below.

What type of graph will you draw?
What will be your independent variable (x axis)?
What will be your dependent variable (y axis)?

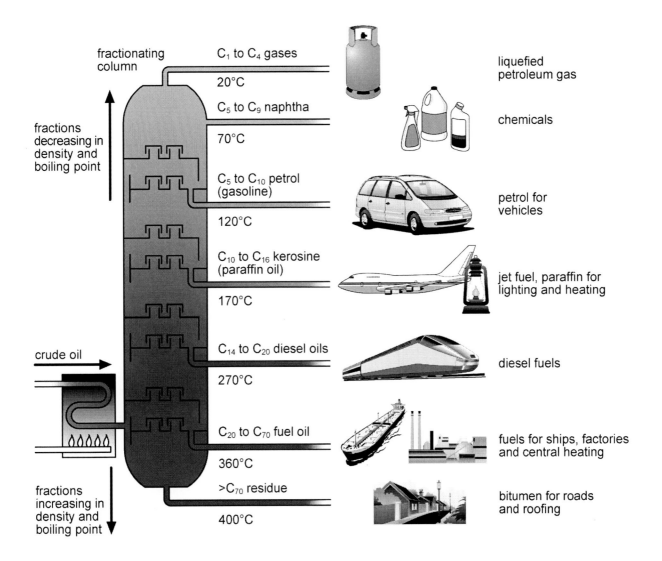

WORLD OIL RESERVES: THE FUTURE?

27

KS4 NATIONAL CURRICULUM HSW link

3. *Communication skills*
   a. recall, analyse, interpret, apply and question scientific information or ideas

RESOURCES:
Task Sheet 27, enough for one per learner or one between two learners.

TIME:
10 minutes for the activity, 5 minutes for class discussion.

NOTES

- This activity is most suitable for use as an extended Starter, or a Main or Homework Activity.
- Learners must be aware of these key words/concepts before attempting the task: graph analysis; identifying trends and making predictions.

SUGGESTED ANSWERS

A. Identify the independent variable and the dependent variable.
   Independent variable: years.
   Dependent variable: giga barrels per annum.

B. Describe the trend of oil production.
   Oil production increased rapidly and exponentially until the mid 1970s. Thereafter the production increased slowly.

C. Describe the overall trend of past oil discovery.
   Oil discovery increased from the 1930s until it peaked during the 1970s. Since then, oil discovery has declined steadily.

D. Is there a relationship between discovery and production?
   As oil discovery increased, the production of oil increased, with a lag phase of about 10 years. However, this relationship did not continue after oil discovery peaked during the 1970s and, since the 1980s, oil production has overtaken oil discovery.

E. Looking at the graph, what do you think the trend of future discovery will be?
   Future discovery rates will decline even further – predictions suggest that oil reserves will only last for another 40 years or so, depending on demand.

EXTENSION SUGGESTION

What alternatives are there to using oil as a fuel to power cars and generate electricity.

WORLD OIL RESERVES: THE FUTURE?

27

It is important to know how much oil is left and so using a graph can help scientists make predictions. Study the graph below:

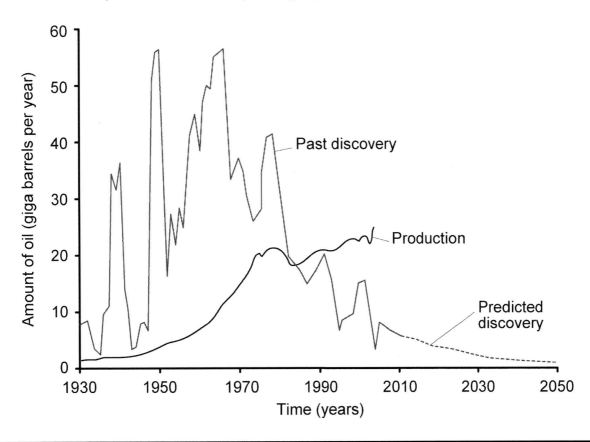

TASK

Use the graph to answer these questions:

A. Identify the independent variable and the dependent variable.

B. Describe the trend of oil production.

C. Describe the overall trend of past oil discovery.

D. Is there a relationship between discovery and production?

E. Looking at the graph, what do you think the trend of future discovery will be?

DESALINATION OF SEA WATER: BENEFITS, DRAWBACKS AND RISKS

28

KS4 NATIONAL CURRICULUM HSW LINK

4. *Applications and implications of science*
   a. about the use of contemporary scientific and technological developments and their benefits, drawbacks and risks

RESOURCES:
Task Sheet 28, cut into 9 cards (1 instruction and 8 sorting cards), enough for one set between two or four learners.

TIME:
10 minutes for the activity, 10 minutes for class discussion.

NOTES

- This task is suitable for use as an extended Starter or Plenary, or a Main Activity.
- Learners must be aware of these key words/concepts before attempting the task: benefit, drawback and risk; desalination.

SUGGESTED ANSWERS

Below is the intention of each statement: Benefit (B), Drawback (D) or Risk (R).

- Desalination plants can destroy coastal habitats such as wetlands and floodplains. *(R)*
- Desalination of sea water will provide a 'drought-proof' supply of water. *(B)*
- Sea water will always be available. *(B)*
- Desalination results in large emissions of carbon dioxide. *(D)*
- Desalination of sea water uses a lot of energy. *(D)*
- The waste product of desalination, concentrated brine, can damage marine ecosystems unless it is carefully mixed when returned to the sea. *(R)*
- Water produced from distilled sea water costs more per litre than water produced from other natural sources. *(D)*
- Desalination of sea water will provide water when other sources of water are dependent on the climate. *(B)*

EXTENSION SUGGESTION

List any other benefits, drawbacks or risks you can think of.

DESALINATION OF SEA WATER: BENEFITS, DRAWBACKS AND RISKS

28

TASK

Discuss each statement card and decide whether it is a benefit, drawback or a risk of desalination of sea water.

- A **benefit** is something that generally has a good effect on people.
- A **drawback** is something that is a hindrance or is the 'downside'.
- A **risk** is a possible danger or source of harm.

Desalination plants can destroy coastal habitats such as wetlands and floodplains.	Desalination of sea water will provide a 'drought-proof' supply of water.
Sea water will always be available.	Desalination results in large emissions of carbon dioxide.
Desalination of sea water uses a lot of energy.	The waste product of desalination, concentrated brine, can damage marine ecosystems unless it is carefully mixed when returned to the sea.
Water produced from distilled sea water costs more per litre than water produced from other natural sources.	Desalination of sea water will provide water when other sources of water are dependent on the climate.

DESALINATION OF SEA WATER IN AUSTRALIA: SOCIAL, ECONOMIC & ENVIRONMENTAL ISSUES

29

KS4 NATIONAL CURRICULUM HSW LINK

4. *Applications and implications of science*
   b. to consider how and why decisions about science and technology are made, including those that raise ethical issues, and about the social, economic and environmental effects of such decisions

RESOURCES:
Task Sheet 29, cut up into 7 cards (1 instruction and 6 sorting cards), enough for one set between two to four learners.

TIME:
10 minutes for the activity, 10 minutes for class discussion.

NOTES

- This task is most suitable for use as a Starter, Main Activity or Plenary.
- Learners must be aware of these key words/concepts before attempting the task: social, economic and environmental; desalination.

SUGGESTED ANSWERS

Below is the intention of each statement: Social (S), Economic (Ec) or Environmental (Env).

- "Desalination plants require huge amounts of energy to distil water from sea water – fossil fuels provide most of this energy." *(Ec)*
  Simon, Environmental Campaigner
- "We don't just need water for drinking - people need water to improve the quality of their lives, for example, water for watering their gardens, golf courses, parks and open spaces." *(S)*
  Shenaz, Local Government Representative
- "Our ranches and farms need fresh water. If our traditional sources of water dry up, we won't be able to farm any more." **Bruce, Farmer** *(S/Ec)*
- "Desalination is a very expensive way to produce fresh water. It would be better to educate people to be less wasteful with their water use." *(S/Ec)*
  Janice, Teacher
- "We are concerned that the levels of salt being returned to the sea will damage delicate marine ecosystems." *(Env)*
  Kylie, Marine Conservation Group
- "Sea water is an excellent source of fresh water because, unlike some other natural water sources, the sea will never run dry!" *(Env)*
  Shane, Desalination Plant CEO

EXTENSION SUGGESTION

Design a method to desalinate salt water in the laboratory.

DESALINATION OF SEA WATER IN AUSTRALIA: SOCIAL, ECONOMIC & ENVIRONMENTAL ISSUES

29

TASK

Sea water contains about 25g of sodium chloride (common salt) per litre. If the salt can be removed, the resulting clean, fresh water could be used by countries that lack other sources of fresh drinking water.

Discuss each statement card and decide whether it is a social, economic or environmental issue about distilling fresh water from sea water.

- **Social**: To do with people's lives and the effect on running a society.
- **Economic**: To do with money, either making money or keeping costs down.
- **Environmental**: To do with keeping our environment unpolluted.

"Desalination plants require huge amounts of energy to distil water from sea water – fossil fuels provide most of this energy." **Simon, Environmental Campaigner**	"We don't just need water for drinking – people need water to improve the quality of their lives, for example, water for watering their gardens, golf courses, parks and open spaces." **Shenaz, Local Government Representative**
"Our ranches and farms need fresh water. If our traditional sources of water dry up, we won't be able to farm any more." **Bruce, Farmer**	"Desalination is a very expensive way to produce fresh water. It would be better to educate people to be less wasteful with their water use." **Janice, Teacher**
"We are concerned that the levels of salt being returned to the sea will damage delicate marine ecosystems." **Kylie, Marine Conservation Group**	"Sea water is an excellent source of fresh water because, unlike some other natural water sources, the sea will never run dry!" **Shane, Desalination Plant CEO**

NATURAL RESOURCES
QUARRYING: SOCIAL, ECONOMIC AND ENVIRONMENTAL ISSUES

30

KS4 NATIONAL CURRICULUM HSW LINK

4. *Applications and implications of science*
   b. to consider how and why decisions about science and technology are made, including those that raise ethical issues, and about the social, economic and environmental effects of such decisions

RESOURCES:
Task Sheet 30, cut up into 7 cards (1 instruction and 6 sorting cards), one set between two to four learners.

TIME:
10 minutes for the activity, 10 minutes for class discussion.

NOTES

- This task is most suitable for use as a Starter, Main Activity or Plenary.
- Learners must be aware of these key words/concepts before attempting the task: social, economic and environmental.

SUGGESTED ANSWERS

Below is the intention of each statement: Social (S), Economic (Ec) or Environmental (Env).

- Materials from quarries are essential to us all. We need the products for building homes, roads, railways and airports. Also, quarry products are important in the production of glass, steel, paper, plastics and cosmetics. *(S/Ec)*
- Quarrying involves removing large amounts of rock or other materials from the landscape. This destroys natural habitats. *(Env)*
- Some quarries require extraction processes on site, which use or produce harmful materials that can find their way into the waterways, e.g. the extraction of some metals. *(Env)*
- Quarried products are very valuable. Large quarrying companies in the UK have an annual turnover of hundreds of millions of pounds. *(Ec)*
- Disused quarries can become wonderful habitats for animals and plants, some become Sites of Special Scientific Interest (SSSI), e.g. Kilmersdon Road Quarry SSSI, Somerset. *(S/Env)*
- Sometimes sites that are proposed for quarrying have historical importance, such as historical battle grounds, or are habitats of rare or endangered species. This causes people to raise objections to the development of such quarries. *(S)*

EXTENSION SUGGESTION

Find out about local quarries, used or disused.

QUARRYING: SOCIAL, ECONOMIC AND ENVIRONMENTAL ISSUES

30

TASK

Discuss each statement card and decide whether it is a social, economic or environmental issue about quarrying.

- **Social**: To do with people's lives and the effect on running a society.
- **Economic**: To do with money, either making money or keeping costs down.
- **Environmental**: To do with keeping our environment unpolluted.

Materials from quarries are essential to us all. We need the products for building homes, roads, railways and airports. Also, quarry products are important in the production of glass, steel, paper, plastics and cosmetics.	Quarried products are very valuable. Large quarrying companies in the UK have an annual turnover of hundreds of millions of pounds.
Quarrying involves removing large amounts of rock or other materials from the landscape. This destroys natural habitats.	Disused quarries can become wonderful habitats for animals and plants, some become Sites of Special Scientific Interest (SSSI), e.g. Kilmersdon Road Quarry SSSI, Somerset.
Some quarries require extraction processes on site, which use or produce harmful materials that can find their way into the waterways, e.g. the extraction of some metals.	Sometimes sites that are proposed for quarrying have historical importance, such as historical battle grounds, or are habitats of rare or endangered species. This causes people to raise objections to the development of such quarries.

MATERIAL PROPERTIES
ARTIFICIAL FERTILISERS

KS4 NATIONAL CURRICULUM HSW LINK

1. *Data, evidence, theories and explanations*
   a. how scientific data can be collected and analysed
   b. how interpretation of data, using creative thought, provides evidence to test ideas and develop theories

RESOURCES:
Task Sheet 31, enough for one per learner or one between two.

TIME:
10 minutes for the activity, 5 minutes for class discussion.

NOTES

- This task is most suitable for use as a Starter, Main Activity or Plenary.
- Teachers may need to discuss the graph with lower ability learners before they attempt the task.
- Learners must be aware of these key words/concepts before attempting the task: graph interpretation.

SUGGESTED ANSWERS

A. Identify the independent and dependent variables.
   Independent variable: time. Dependent variable: mass of artificial fertiliser.
B. What units could you use to measure the amount of artificial fertiliser used per year? *Megatonnes*
C. Describe in words what the graph shows is happening to the amount of artificial fertiliser used.
   The graph shows a steady increase from 1950 to 1980 but then it slows and begins to level out.
D. How does the amount of artificial fertiliser used in the UK compare with the amount used globally over the past 50 years?
   Fertiliser use in the UK has levelled out and has begun to decrease. Globally, levels of fertilisers have levelled out but have not fallen.

EXTENSION SUGGESTION

Suggest why the patterns of artificial fertiliser usage are different in the UK and globally.

This graph shows the amount of artificial fertilisers used each year globally since 1950.

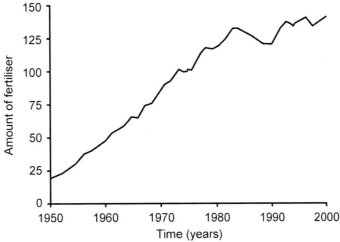

TASK

A. Identify the independent and dependent variables.

B. What units could you use to measure the amount of artificial fertiliser used per year?

C. Describe in words what the graph shows is happening to the amount of artificial fertiliser used.

A graph for the amount of artificial fertiliser used in the UK over the same time period is shown below.

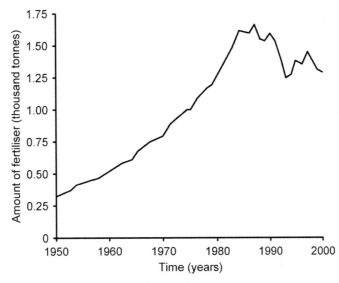

D. How does the amount of artificial fertiliser used in the UK compare with the amount used globally over the past 50 years?

DO FOOD COLOURINGS AFFECT BEHAVIOUR?

32

KS4 NATIONAL CURRICULUM HSW LINK

1. *Data, evidence, theories and explanations*
   d. evaluate methods of collection of data and consider their validity and reliability as evidence

RESOURCES:
Task Sheet 32, enough for one between two learners.

TIME:
15 minutes for the activity, 10 minutes for class discussion.

NOTES

- This task is most suitable for use as an extended Starter, Main Activity or Plenary.
- Learners must be aware of these key words/concepts before attempting the task: placebo; double-blind trial; synthetic food colourings.
- Learners with low literacy skills will need support with reading the report.
- The text was summarised from the original published paper: http://www.ncbi.nlm.nih.gov/pubmed/7965420

SUGGESTED ANSWERS

The conclusions that are supported by the study are:
- Some children are sensitive to synthetic food colourings such as tartrazine.
- Some children's behaviour is affected by the food colouring tartrazine.
- Ingesting tartrazine is correlated with irritability, restlessness and sleep disturbance in some children.

EXTENSION SUGGESTION

There is an additional task at the end of the Task Sheet.

DO FOOD COLOURINGS AFFECT BEHAVIOUR?

32

Read the summary of this scientific study.

Australian scientists wanted to find out whether there is an association between the ingestion of synthetic food colourings and hyperactive children.

The experiment was carried out on 200 children. They were fed on a diet free of synthetic food colourings for six weeks. Using a behaviour questionnaire, 150 parents reported that their children's behaviour improved after the 6 weeks on the diet.

Then 34 children who were thought to be 'sensitive' to artificial colourings and 20 children in a control group were studied. The children were all on the synthetic colouring free diet. A 21-day double-blind, placebo-controlled trial was carried out. Each child was given either a placebo, or a dose of the synthetic food colouring tartrazine, each morning. The child's parents filled out a behaviour rating at the end of each day.

The results were 22 of the 34 'sensitive' children and 2 of the control group showed significant changes in behaviour. These children were all reported to be irritable, sleepless and restless.

TASK

From this study, discuss which of these conclusions are supported:

- All food colourings cause bad behaviour in children.
- All children's behaviour is affected by food colourings.
- Some children are sensitive to synthetic food colourings, such as tartrazine.
- Some children's behaviour is affected by the food colouring tartrazine.
- Ingesting tartrazine causes irritability, restlessness and sleep disturbance in some children.
- Ingesting tartrazine is correlated with irritability, restlessness and sleep disturbance in some children.

EXTENSION

What further studies would need to be carried out to support some of these conclusions?

MATERIAL PROPERTIES
WHICH WASHING POWDER WASHES CLEANEST?

KS4 NATIONAL CURRICULUM HSW LINK

2. *Practical and enquiry skills*
   a. plan to test a scientific idea, answer a scientific question or solve a scientific problem
   c. work accurately and safely, individually and with others, when collecting first-hand data

RESOURCES:
Task Sheet 33, enough for one between one or two learners.

TIME:
15 minutes for the activity, 10 minutes for class discussion.

NOTES

- This task is most suitable for use as an extended Starter, Main Activity or Plenary. It could also be set as a Homework Activity.
- Learners must be aware of these key words/concepts before attempting the task: planning an investigation; valid, reliable, accurate.

SUGGESTED ANSWERS

Prediction with a scientific reason:
Traditional washing powders remove more stains than environmentally friendly powders because they contain more effective chemicals.

Bullet point method (how you will do the experiment). Include:
- *Same volume of water and temperature of water.*
- *Same amount of time being washed.*
- *Same mass of powder.*
- *Same treatment, e.g. agitation, stirring, etc.*
- *Same type of stain and length of time stain is left on material before washing.*

Safety considerations: *hot water, chemicals in powder could be harmful.*

State how you will ensure that the experiment is:
- valid – *comparing the two powders, controlling other variables.*
- reliable – *repeat readings.*
- accurate – *how stain removal is measured.*

EXTENSION SUGGESTION

Draw a table for the results.

WHICH WASHING POWDER WASHES CLEANEST?

Traditional washing powders clean clothes very well, they make the clothes smell fresh and keep the colours looking bright. However, they contain many chemicals which are harmful to the environment.

Environmentally friendly washing powders are available that are completely biodegradable and have very little negative impact on the environment. However, these washing powders are generally more expensive than traditional washing powders and some consumers say that they don't wash clothes as effectively.

TASK

Your task is to plan an investigation to test how well a traditional washing powder washes clothes compared with an environmentally friendly brand.

Include:

- Prediction with a scientific reason.
- Bullet point method (how you will do the experiment).
- Safety considerations.

State how you will ensure that the experiment is:

- valid
- reliable
- accurate

THERMOPLASTIC OR THERMOSETTING PLASTIC?

34

KS4 NATIONAL CURRICULUM HSW LINK

2. Practical and enquiry skills

    a. plan to test a scientific idea, answer a scientific question or solve a scientific problem

RESOURCES:
Task Sheet 34, enough for one per learner or one between two.

TIME:
10 minutes for the activity, 5 minutes for class discussion.

NOTES

- This task is most suitable for use as a Starter, Main Activity or Plenary.
- Learners must be aware of these key words/concepts before attempting the task: using secondary data; thermosetting; thermoplastic.

SUGGESTED ANSWERS

Plastic tested	Appearance and flexibility	Reaction to heating	Thermoplastic or thermosetting
Bit of old phone	Black, shiny, brittle and not flexible	Charred a bit at edges where heated	*Thermosetting*
Piece of guttering	Brown, flexible	Went soft and gloopy and dripped	*Thermoplastic*
Carrier bag	Orange, very thin, stretchy	Sort of shrank into nothing – melted?	*Thermoplastic*
Milk bottle crate	Green, hard, rigid	Went black at the edges	*Thermosetting*
Milk carton	Sort of clear, flexible and easy to squash	Changed into stringy, stretchy stuff, made a mess	*Thermoplastic*
Camera film	Black, flexible, strong	Melted very quickly	*Thermoplastic*
Piece of silky material	Shiny, soft and silky, pale blue	Sort of shrank to nothing like plastic bag	*Thermoplastic*
Child's bowl	White with animal print, bit scratched, very hard	Cracked, went black	*Thermosetting*

EXTENSION SUGGESTION

What other properties could the students have tested?

The information below describes the properties of two important types of plastic – thermosetting and thermoplastic.

Thermosetting	Thermoplastic
• Heat resistant • Can be melted and moulded only once • Chars (burns) • Cracks or disintegrates • Tends to be brittle and inflexible when heated a second time • Lines of molecules with many cross links	• Melts between temperatures of 65 – 200°C • Can be reheated and remoulded • Fully recyclable • Strong but flexible • Made up of lines of molecules with few cross links

TASK

Some students tested samples of plastics to identify which ones were thermosetting plastics and which ones were thermoplastic (see the table below). Use the information above to decide which of the tested plastics are thermoplastic and which are thermosetting.

Plastic tested	Appearance and flexibility	Reaction to heating	Thermoplastic or thermosetting
Bit of old phone	Black, shiny, brittle and not flexible	Charred a bit at edges where heated	
Piece of guttering	Brown, flexible	Went soft and gloopy and dripped	
Carrier bag	Orange, very thin, stretchy	Sort of shrank into nothing – melted?	
Milk bottle crate	Green, hard, rigid	Went black at the edges	
Milk carton	Sort of clear, flexible and easy to squash	Changed into stringy, stretchy stuff, made a mess	
Camera film	Black, flexible, strong	Melted very quickly	
Piece of silky material	Shiny, soft and silky, pale blue	Sort of shrank to nothing like plastic bag	
Child's bowl	White with animal print, bit scratched, very hard	Cracked, went black	

WHICH PLASTIC CARRIER BAG IS BEST?

35

KS4 NATIONAL CURRICULUM HSW LINK

3. *Communication skills*
   a. recall, analyse, interpret, apply and question scientific information or ideas
   b. use both qualitative and quantitative approaches
   c. present information, develop an argument and draw a conclusion, using scientific, technical and mathematical language, conventions and symbols, and ICT tools

RESOURCES:
Task Sheet 35, enough for one between two learners.

TIME:
15 minutes for the activity, 10 minutes for class discussion.

NOTES

- This task is most suitable for use as an extended Starter, Main Activity or Plenary. It could also be set as a Homework Activity.
- Learners must be aware of these key words/concepts before attempting the task: planning an investigation.

SUGGESTED ANSWERS

A. What are the problems with comparing these claims?
   The claims are not easily comparable.
   There is a difference between 'average' and 'up to'.
B. What could you ask the people who tested these bags about their methods?
   How many times did you repeat the experiment?
   What was the range of the results?
   What size can did you use? Etc.
C. Come up with a test that would compare these bags fairly and reliably. Include:
   Use same size masses to add to the bags.
   Time how long the bags can hold each mass.
   Support the handles of the bags in the same way, e.g. clip, hook, threaded onto a bar.

EXTENSION SUGGESTION

Do the experiment.

WHICH PLASTIC CARRIER BAG IS BEST?

35

Supermarkets have recently tried to improve their plastic bags by making them stronger and reusable.

Read these statements:

COSTE	DASA	Burysains
Our bags can hold an average of 10 cans of baked beans for 20 minutes.	Our bags can hold six 2 litre bottles of cola for up to 15 minutes.	Our bags are biodegradable and can cope with an average shop of 6kg.

TASK

Discuss the following:

A. What are the problems with comparing these claims?

B. What could you ask the people who tested these bags about their methods?

C. Come up with a test that would compare these bags fairly and reliably.

KS4 NATIONAL CURRICULUM HSW link

3. *Communication skills*
   c. present information, develop an argument and draw a conclusion, using scientific, technical and mathematical language, conventions and symbols, and ICT tools

RESOURCES:
Task Sheet 36, cut up into 25 cards (1 instruction and 24 sorting cards), one set between two to four learners.

TIME:
15 minutes for the activity, 10 minutes for class discussion.

NOTES

- This task is most suitable for use as an extended Starter, Main Activity or Plenary. It could also be set as a Homework Activity.
- Learners must be aware of these key words/concepts before attempting the task: scale; nanotechnology.

SUGGESTED ANSWERS

Unit & symbol	Example	Size relative to metres
kilometres km	Local town	1×10^{3}
metres m	Your desk	1×10^{1}
centimetres cm	Your hand	1×10^{-2}
millimetres mm	Length of fingernails	1×10^{-3}
micrometres μm	Length of skin cells	1×10^{-6}
nanometres nm	Width of molecules like haemoglobin and DNA	1×10^{-9}
picometres pm	Width of individual atoms	1×10^{-12}

EXTENSION SUGGESTION

Put the cards in order of size.

SCALE: BIG TO VERY SMALL

TASK

Match the unit and symbol to the example and size relative to metres.

Unit & symbol	Example	Size relative to metres
metres m	Length of fingernails	1×10^{-12}
centimetres cm	Your desk	1×10^{1}
millimetres mm	Your hand	1×10^{-2}
micrometres μm	Width of molecules like haemoglobin and DNA	1×10^{-3}
picometres pm	Length of skin cells	1×10^{-6}
nanometres nm	Width of individual atoms	1×10^{-9}
kilometres km	Local town	1×10^{3}

MATERIAL PROPERTIES
WASHING POWDER PROPERTIES: ESSENTIAL OR DESIRABLE

37

KS4 NATIONAL CURRICULUM HSW LINK

4. *Applications and implications of science*
   a. about the use of contemporary scientific and technological developments and their benefits, drawbacks and risks
   b. to consider how and why decisions about science and technology are made, including those that raise ethical issues, and about the social, economic and environmental effects of such decisions

RESOURCES:
Task Sheet 37, cut up into 10 cards (1 instruction and 9 sorting cards), enough for one set between two to four learners.

TIME:
10 minutes for the activity, 10 minutes for class discussion.

NOTES

- This task is most suitable for use as a Starter, Main Activity or Plenary.
- Learners must be aware of these key words/concepts before attempting the task: essential; desirable; water softener.

SUGGESTED ANSWERS

Intention of answers (E = Essential, D = Desirable)
- Surfactant to improve the wetting ability of water.   (E)
- Does not damage clothing fibres.   (E)
- 'Builders', e.g. zeolites, to act as water softeners.   (E)
- Cleans clothes at a range of temperatures.   (D)
- Does not irritate sensitive skin.   (D)
- 100% biodegradable.   (D)
- Optical brighteners to make whites look 'whiter'.   (D)
- Gives laundry a pleasant smell.   (D)
- 'Fillers', e.g. sodium sulphate, to make the powder flow more easily.   (D)

EXTENSION SUGGESTION

Suggest additional qualities it is desirable for a washing powder to have.

WASHING POWDER PROPERTIES: ESSENTIAL OR DESIRABLE

37

TASK

Discuss each statement card and decide whether it is an essential property of a washing powder or a just a desirable quality.

- **Essential**: Required to make the product effective.
- **Desirable**: A preferred property but not essential.

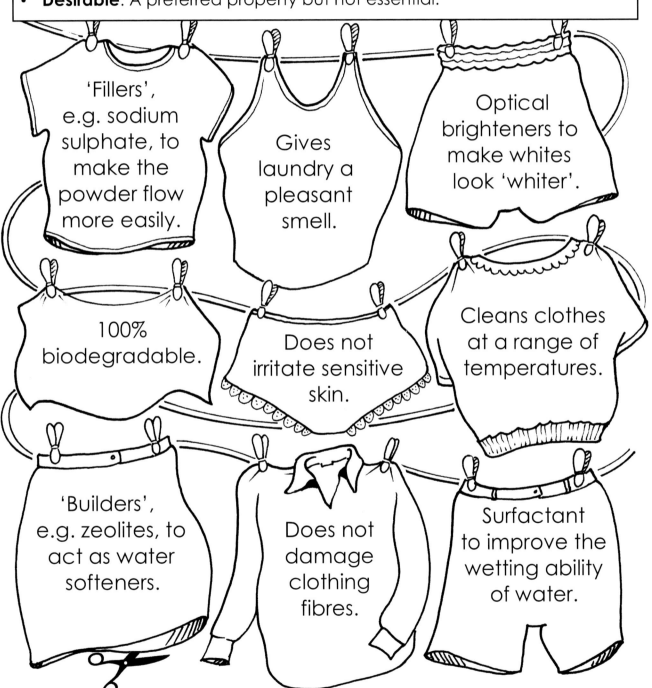

'Fillers', e.g. sodium sulphate, to make the powder flow more easily.

Gives laundry a pleasant smell.

Optical brighteners to make whites look 'whiter'.

100% biodegradable.

Does not irritate sensitive skin.

Cleans clothes at a range of temperatures.

'Builders', e.g. zeolites, to act as water softeners.

Does not damage clothing fibres.

Surfactant to improve the wetting ability of water.

NANOTECHNOLOGY: BENEFITS, DRAWBACKS AND RISKS

38

KS4 NATIONAL CURRICULUM HSW LINK

4. *Applications and implications of science*
   a. about the use of contemporary scientific and technological developments and their benefits, drawbacks and risks

RESOURCES:
Task Sheet 38, cut up into 9 cards (1 instruction and 8 sorting cards), enough for one set between two to four learners.

TIME:
10 minutes for the activity, 10 minutes for class discussion.

NOTES

- This task is suitable for use as an extended Starter or Plenary, or a Main Activity.
- Learners must be aware of these key words/concepts before attempting the task: benefit, drawback and risk.

SUGGESTED ANSWERS

Below is the intention of each statement: Benefit (B), Drawback (D) or Risk (R).

- Nanotechnology could make microscopic robots to deliver drugs directly to specific parts of the body, reducing side effects. *(B)*
- Some newspapers have run 'scare-stories' that nanotechnology could create a 'grey-goo' that could take over the world! [*Drawback that people think this!*] *(D)*
- It is difficult to know exactly how some nanotechnology will work. *(D)*
- Terrorists could use nanotechnology to make chemical or biological weapons. *(R)*
- Nanotechnology could make intelligent clothes that react to the wearer's needs, such as let out more sweat, hold in more heat and so on. *(B)*
- Nanotubes could be used as mini fuel cells to run mobile cells on methane. *(B)*
- Nanotechnology is already used in many building materials, such as self-cleaning glass and concrete. *(B)*
- Nanotechnology is used in sun block creams, using nanoparticles to reflect UV radiation. These creams are invisible on the skin. *(B)*

EXTENSION SUGGESTION

Are these statements a balanced representation of views?
Look up the idea of 'grey-goo'. What is it? Is it a real problem?

NANOTECHNOLOGY: BENEFITS, DRAWBACKS AND RISKS

38

TASK

Discuss each statement card and decide whether it is a benefit, drawback or a risk of using nanotechnology.

- A **benefit** is something that generally has a good effect on people.
- A **drawback** is something that is a hindrance or is the 'downside'.
- A **risk** is a possible danger or source of harm.

Nanotechnology could make microscopic robots to deliver drugs directly to specific parts of the body, reducing side effects.	Some newspapers have run 'scare-stories' that nanotechnology could create a 'grey-goo' that could take over the world!
It is difficult to know exactly how some nanotechnology works.	Terrorists could use nanotechnology to make chemical or biological weapons.
Nanotechnology could make intelligent clothes that react to the wearer's needs, such as let out more sweat, hold in more heat and so on.	Nanotubes could be used as mini fuel cells to run mobile cells on methane.
Nanotechnology is already used in many building materials, such as self-cleaning glass and concrete.	Nanotechnology is used in sun block creams, using nanoparticles to reflect UV radiation. These creams are invisible on the skin.

ARTIFICIAL FERTILISERS: SOCIAL, ECONOMIC AND ENVIRONMENTAL ISSUES

39

KS4 NATIONAL CURRICULUM HSW LINK

4. *Applications and implications of science*
   b. to consider how and why decisions about science and technology are made, including those that raise ethical issues, and about the social, economic and environmental effects of such decisions

RESOURCES:
Task Sheet 39, cut up into 7 cards (1 instruction and 6 sorting cards), one set between two to four learners.

TIME:
10 minutes for the activity, 10 minutes for class discussion.

NOTES

- This task is most suitable for use as a Starter, Main Activity or Plenary.
- Learners must be aware of these key words/concepts before attempting the task: social, economic and environmental.

SUGGESTED ANSWERS

Below is the intention of each statement: Social (S), Economic (Ec) or Environmental (Env).

- "The use of artificial fertilisers has allowed yields to increase so food has become cheaper – this can only be a good thing for the consumer." **Alice, Supermarket Manager** *(Ec)*
- "At the present time, artificial fertilisers are made in an unsustainable way – for example, the Haber-Bosch process used to fix nitrogen uses a lot of energy." **Joe, Environmental Campaigner** *(Env)*
- "Excessive use of artificial fertilisers can lead to eutrophication in water systems." **Mandy, Water Authority Spokesperson** *(Env)*
- "Artificial fertilisers have enabled me to grow the same crop in the same field year after year, and my yields have increased." **Tony, Farmer** *(S/Env)*
- "There are no alternatives to artificial fertilisers on a global scale if sufficient food is to be grown." **Jeremy, Agrochemical Industry Spokeperson** *(S)*
- "I don't like the smell when they put fertilisers on the fields – but it smells better than muck spreading!" **Laura, Primary School Student** *(S/Env)*

EXTENSION SUGGESTION

Suggest alternatives to using artificial fertilisers.

ARTIFICIAL FERTILISERS: SOCIAL, ECONOMIC AND ENVIRONMENTAL ISSUES

39

TASK

Discuss each statement card and decide whether it is a social, economic or environmental issue about using artificial fertilisers.

- **Social**: To do with people's lives and the effect on running a society.
- **Economic**: To do with money, either making money or keeping costs down.
- **Environmental**: To do with keeping our environment unpolluted.

"The use of artificial fertilisers has allowed yields to increase so food has become cheaper – this can only be a good thing for the consumer." **Alice, Supermarket Manager**	"At the present time, artificial fertilisers are made in an unsustainable way – for example, the Haber-Bosch process used to fix nitrogen uses a lot of energy." **Joe, Environmental Campaigner**
"Excessive use of artificial fertilisers can lead to eutrophication in water systems." **Mandy, Water Authority Spokesperson**	"Artificial fertilisers have enabled me to grow the same crop in the same field year after year, and my yields have increased." **Tony, Farmer**
"There are no alternatives to artificial fertilisers on a global scale if sufficient food is to be grown." **Jeremy, Agrochemical Industry Spokeperson**	"I don't like the smell when they put fertilisers on the fields – but it smells better than muck spreading!" **Laura, Primary School Student**

PLASTIC BAGS: BENEFITS, DRAWBACKS AND RISKS

40

KS4 NATIONAL CURRICULUM HSW LINK

4. *Applications and implications of science*
   a. about the use of contemporary scientific and technological developments and their benefits, drawbacks and risks

RESOURCES:
Task Sheet 40, cut up into 9 cards (1 instruction and 8 sorting cards), one set between two to four learners.

TIME:
10 minutes for the activity, 10 minutes for class discussion.

NOTES

- This task is suitable for use as an extended Starter or Plenary, or a Main Activity.
- Learners must be aware of these key words/concepts before attempting the task: benefit, drawback and risk.

SUGGESTED ANSWERS

Below is the intention of each statement: Benefit (B), Drawback (D) or Risk (R).

- Compared to paper shopping bags, plastic shopping bags require less energy to produce, transport and recycle. *(B)*
- Plastic shopping bags are durable, low cost and strong. *(B)*
- Plastic bags are not biodegradable. *(D)*
- Plastic shopping bags are made from oil-based chemicals, which are a non-renewable resource. *(D)*
- If plastic shopping bags are not disposed of carefully, they can block drains and harm wildlife. *(R)*
- Although plastic bags are recyclable, very few actually are recycled. *(D)*
- Plastic shopping bags are reusable. *(B)*
- A number of infants and young children accidentally suffocate in plastic bags each year. *(R)*

EXTENSION SUGGESTION

List any other benefits, drawbacks or risks you can think of that relate to the use of plastic bags.

PLASTIC BAGS: BENEFITS, DRAWBACKS AND RISKS

TASK

Discuss each statement card and decide whether it is a benefit, drawback or a risk of using plastic carrier bags.

- A **benefit** is something that generally has a good effect on people.
- A **drawback** is something that is a hindrance or is the 'downside'.
- A **risk** is a possible danger or source of harm.

Compared to paper shopping bags, plastic shopping bags require less energy to produce, transport and recycle.	Plastic shopping bags are durable, low cost and strong.
Plastic bags are not biodegradable.	Plastic shopping bags are made from oil-based chemicals, which are a non-renewable resource.
If plastic shopping bags are not disposed of carefully, they can block drains and harm wildlife.	Although plastic bags are recyclable, very few actually are recycled.
Plastic shopping bags are reusable.	A number of infants and young children accidentally suffocate in plastic bags each year.

WHY DID THE OXIDATION CATASTROPHE HAPPEN?

41

KS4 NATIONAL CURRICULUM HSW LINK

1. *Data, evidence, theories and explanations*
   a. how scientific data can be collected and analysed
   b. how interpretation of data, using creative thought, provides evidence to test ideas and develop theories
   c. how explanations of many phenomena can be developed using scientific theories, models and ideas
   d. that there are some questions that science cannot currently answer, and some that science cannot address

RESOURCES:
Task Sheet 41, enough for one per learner.

TIME:
15 minutes for the activity, 10 minutes for class discussion.

NOTES

- This task is suitable for use as a Main or Homework Activity.
- Learners with reading difficulties will need support either by a peer or by the teacher reading through the text.
- Learners must be aware of these key words/concepts before attempting the task: cyanobacteria; photosynthesis; oxidation of iron; mantle; plumes; computer models; geology.

SUGGESTED ANSWERS

A. What the two scientists agree on.
   There was a small amount of oxygen in the Earth's atmosphere prior to 2.4 billion years ago. Photosynthetic cyanobacteria were present and were producing oxygen before the oxidation event. The catastrophic event resulted in the laying down of 'red beds'.

B. What the two scientists disagree about.
   What caused the dramatic increase in oxygen concentrations: either biological or geological. Where the oxygen originated (either through photosynthesis or deep mantle plumes).

C. The evidence they use in their arguments. *Rock records, fossil records.*

D. Why it is difficult to be sure exactly why the levels of oxygen increased so rapidly. *Incomplete biological and geological records, no one was there at the time.*

EXTENSION SUGGESTION

List any questions you have about the statements.

WHY DID THE OXIDATION CATASTROPHE HAPPEN?

41

Scientists agree that sometime between 2.3 and 2.4 billion years ago, oxygen levels in the Earth's atmosphere increased dramatically. However, because the events took place millions of years ago, it is difficult to know exactly why levels of oxygen increased.

TASK

Read the views of these two scientists. Then complete the task below:

Dr. Hessa Guess

We know the levels of oxygen increased between 2.4 to 2.3 billion years ago because of the presence of iron oxides in fossil soils. These soils are commonly known as 'red beds'. The rapid increase in oxygen levels resulted in the extinction of many anaerobic bacteria. The cause of this 'oxidation catastrophe' was biological.

Cyanobacteria and early plants were photosynthesising and releasing oxygen millions of years before the catastrophic oxygenation of the Earth's atmosphere occurred. Fossils of cyanobacteria have been found dating back to 2.5 billion yeas ago. The cyanobacteria would have lived in shallow pools of water. The fossil records show that numbers of cyanobacteria increased around 2.3 billion years ago and, as they became abundant and widespread, so the levels of oxygen increased. Life on Earth not only adapts to conditions on the planet but also exerts some control over it.

Prof. Ivor Goodideer

Between 2.4 to 2.3 billion years ago, oxygen levels increased dramatically. The fossil record shows that cyanobacteria were present on the Earth's surface some 150 million years before the catastrophic oxidation event. However, the oxygen they were releasing would have reacted with volcanic gases, such as hydrogen, carbon monoxide and methane, and also with huge sinks of reduced iron. Geological rock records can be used to show how levels of oxidised iron change over time. 'Red beds' older than 2.3 billion years are never found.

The increase in levels of oxygen we see at 2.4 to 2.3 billion years ago was caused by geological events. The latest computer model (based on rock records) shows that iron on the sea bed reacted with oxygen to form iron oxides (rusts) which were then subducted and returned to the magma. The oxygen-rich magma was spewed out of deep mantle plumes, adding oxygen to the atmosphere.

Discuss and decide on:
A. What the two scientists agree on.
B. What the two scientists disagree about.
C. The evidence they use in their arguments.
D. Why it is difficult to be sure exactly why the oxygen levels increased so rapidly.

KS4 NATIONAL CURRICULUM HSW LINK

1. *Data, evidence, theories and explanations*
   b. how interpretation of data, using creative thought, provides evidence to test ideas and develop theories
   c. how explanations of many phenomena can be developed using scientific theories, models and ideas
   d. that there are some questions that science cannot currently answer, and some that science cannot address

RESOURCES:
Task Sheet 42, cut into 15 cards (1 instruction and 14 sorting cards), one between two or four learners, or one each.

TIME:
15 minutes for the activity, 10 minutes for class discussion.

NOTES

- This task is suitable for use as a Main or Homework Activity.
- Learners must be aware of these key words/concepts before attempting the task: photosynthesis; fossilisation.

SUGGESTED ANSWERS

4600 million years ago	Early Earth is bombarded by impacts and solar winds, stripping the Earth of its first atmosphere.
4000 million years ago	The first volcanoes erupt, producing heavier gases such as ammonia, carbon dioxide and water vapour. The condensing water vapour forms seas. Nitrogen is released from ammonia due to violent lightning in storms.
2500 million years ago	First bacteria and single celled organisms evolve in the sea and use photosynthesis, releasing oxygen into the atmosphere and taking carbon dioxide in. Seas dissolve carbon dioxide. Bacteria using ammonia release nitrogen.
2000 million years ago	The increasing amount of oxygen from the increasing population of microbes' photosynthesis allows an early ozone layer to form. Numerous bacteria continue to release nitrogen.
1000 million years ago	In the shelter of the ozone layer, UV light hitting the Earth decreases and life begins to flourish in the sea. A lot of carbon dioxide is fixed in sea animals that die to form rock and oil under the oceans.
500 million years ago	The first land plants and animals evolve; photosynthesis is at a high level. Even more carbon dioxide is absorbed by the sea, added to the shells of marine animals and trapped in rocks. Later, tropical forests fossilise to become coal.
0 million years ago	Chemicals released from one of the activities of one species of mammal start to thin the ozone layer. Deforestation and burning of fossil fuels releases carbon dioxide that has been trapped in rocks.

EXTENSION SUGGESTION

How do scientists know this? What evidence is there for these statements?

TASK

Match the descriptions of the processes of Earth with the time and gas levels. Read the gas data carefully to find clues in the descriptions.

2000 million years ago Helium & Hydrogen: negligible Carbon dioxide: 10% (decreasing) Nitrogen: 30% (increasing) Water vapour: some (variable) Oxygen: about 3% (increasing)	The first land plants and animals evolve; photosynthesis is at a high level. Even more carbon dioxide is absorbed by the sea, added to the shells of marine animals and trapped in rocks. Later, tropical forests fossilise to become coal.
4000 million years ago Helium & Hydrogen: negligible Carbon dioxide: 90% (increasing) Nitrogen: 10% (increasing) Water vapour: some (decreasing) Oxygen: None	The increasing amount of oxygen from the increasing population of microbes' photosynthesis allows an early ozone layer to form. Numerous bacteria continue to release nitrogen.
500 million years ago Helium & Hydrogen: negligible Carbon dioxide: 0.5% (decreasing) Nitrogen: 78% (stable) Water vapour: some (variable) Oxygen: about 10% (increasing)	Early Earth is bombarded by impacts and solar winds, stripping the Earth of its first atmosphere.
0 million years ago Helium & Hydrogen: negligible Carbon dioxide: 0.03% (increasing) Nitrogen: 78% (stable) Water vapour: some (variable) Oxygen: 20% (stable)	Chemicals released from one of the activities of one species of mammal start to thin the ozone layer. Deforestation and burning of fossil fuels releases carbon dioxide that has been trapped in rocks.
4600 million years ago Helium: 50% Hydrogen: 50% Carbon dioxide: None Nitrogen: None Water vapour: None Oxygen: None	In the shelter of the ozone layer, UV light hitting the Earth decreases and life begins to flourish in the sea. A lot of carbon dioxide is fixed in sea animals that die to form rock and oil under the oceans.
1000 million years ago Helium & Hydrogen: negligible Carbon dioxide: 5% (decreasing) Nitrogen: 40% (increasing) Water vapour: some (variable) Oxygen: about 5% (increasing)	First bacteria and single celled organisms evolve in the sea and use photosynthesis, releasing oxygen into the atmosphere and taking carbon dioxide in. Seas dissolve carbon dioxide. Bacteria using ammonia release nitrogen.
2500 million years ago Helium & Hydrogen: negligible Carbon dioxide: 50% (decreasing) Nitrogen: 20% (increasing) Water vapour: some (decreasing) Oxygen: about 1% (increasing)	The first volcanoes erupt, producing heavier gases such as ammonia, carbon dioxide and water vapour. The condensing water vapour forms seas. Nitrogen is released from ammonia due to violent lightning in storms.

PREDICTING CARBON DIOXIDE LEVELS

43

KS4 NATIONAL CURRICULUM HSW LINK

1. *Data, evidence, theories and explanations*
   b. how interpretation of data, using creative thought, provides evidence to test ideas and develop theories
   d. that there are some questions that science cannot currently answer, and some that science cannot address

RESOURCES:
Task Sheet 43, one between two learners or one each.

TIME:
15 minutes for the activity, 10 minutes for class discussion.

NOTES

- This task is most suitable for use as an extended Starter, Main or Homework Activity, or Plenary.
- Learners must be aware of these key words/concepts before attempting the task: graph interpretation.

SUGGESTED ANSWERS

A. Describe what the graph shows you. *Concentrations of carbon dioxide remained relatively constant for 800 years before rising after 1800. Initially, the increase was slow but then it speeded up dramatically.*

B. Suggest why concentrations of carbon dioxide in the atmosphere have changed dramatically since 1800. *About 1800 marked the beginning of the industrial revolution.*

C. Predict what will happen to levels of carbon dioxide in the atmosphere if emissions carry on at the current rate. *At the current rate, carbon dioxide concentrations will continue to rise steeply (in 2007, carbon dioxide concentrations reached a record high of 380 ppm, nearly 100 ppm above the average carbon dioxide levels for the past 1000 years). Levels look set to rise, with the continued industrialisation of India and China, for example: China plans to build 500 coal burning power stations within the next few years.*

D. Suggest ways in which carbon dioxide emissions could be reduced. *Examples of ways to reduce carbon emissions include: stop deforestation, reforest cleared land, recycle waste materials, reduce the amount of goods we use, develop renewable energy resources that will match the energy output of fossil fuels, reduce our energy demands, etc.*

EXTENSION SUGGESTION

Suggest possible reasons for the changes in concentrations of carbon dioxide between 1200 and 1400.

This graph shows the amount of carbon dioxide in the atmosphere (in parts per million, ppm) for the past 1000 years.

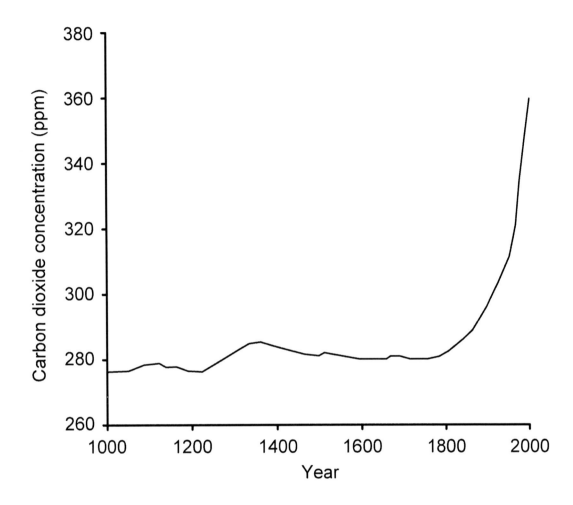

TASK

Discuss the graph in pairs and answer these questions:

A. Describe what the graph shows you.

B. Suggest why concentrations of carbon dioxide in the atmosphere have changed dramatically since 1800.

C. Predict what will happen to levels of carbon dioxide in the atmosphere if emissions carry on at the current rate.

D. Suggest ways in which carbon dioxide emissions could be reduced.

GREENHOUSE GASES: WHICH IS WORST?

44

KS4 NATIONAL CURRICULUM HSW LINK

1. *Data, evidence, theories and explanations*
   a. how scientific data can be collected and analysed
   b. collect data from primary or secondary sources, including using ICT sources and tools
   c. evaluate methods of collection of data and consider their validity and reliability as evidence

RESOURCES:
Task Sheet 44, enough for one per learner.

TIME:
15 minutes for the activity, 5 minutes for class discussion.

NOTES

- This task is most suitable for use as an extended Starter, Main or Homework Activity.
- Learners must be aware of these key words/concepts before attempting the task: presenting data in tables.
- Some learners may require literacy support. Pupils needing support will require the blank table to fill in with headers and GHGs provided.

SUGGESTED ANSWERS

Greenhouse Gas	Concentration pre-1750 (ppm)	Concentration today (ppm)	Global Warming Potential	Contribution to Greenhouse Effect (%)
Carbon dioxide	280	377	1	72.4
Methane	730	1750	21	7.1
Nitrogen dioxide	270	320	300	19.0
CFCs	0	250	5000	1.43

EXTENSION SUGGESTION

How could you find out how long each gas stays in the atmosphere?

GREENHOUSE GASES: WHICH IS WORST? | 44

TASK

Read the text below and put the information into a table that shows the data clearly. Then decide if the author's conclusion is supported by the data.

There are four main gases that contribute to the Greenhouse Effect: carbon dioxide, methane, nitrogen dioxide and chlorofluorocarbons (CFCs).

Before the industrial revolution, the concentration of these greenhouse gases was lower than it is now. Carbon dioxide was about 280 ppm (parts per million) before 1750, methane about 730 ppm and nitrogen dioxide about 270 ppm. There were no CFCs in the atmosphere pre-1750.

Since the industrial revolution, the concentrations of all these gases in the atmosphere have increased. However, not all the gases contribute equally to the Greenhouse Effect.

Carbon dioxide is usually blamed for causing the Greenhouse Effect. Its concentration is currently at 377 ppm. However, if you take the same amount of methane and carbon dioxide, the methane has a Global Warming Potential (GWP) of 21 times that of carbon dioxide. Nitrogen dioxide has a GWP of 300 and CFCs a whopping 5000.

So, not only do you have to consider the types of gas that are being produced, but also their Global Warming Potential. From this, scientists have estimated the contribution of each gas to the Greenhouse Effect. Carbon dioxide is estimated to contribute 72.4% to the Greenhouse Effect, whereas methane, which is currently at a concentration of 1750 ppm, contributes only 7.1%, nitrogen dioxide (at 320 ppm) 19% and CFCs (at 250 ppm) only 1.43%.

Although these figures still point to carbon dioxide being the greatest contributor, despite its low GWP, the data does not take into account how long these gases stay in the atmosphere.

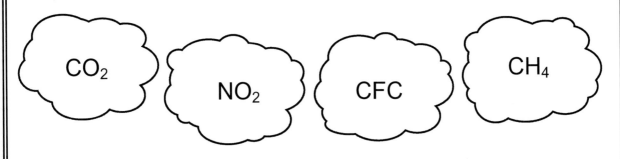

ACID RAIN RESEARCH

KS4 NATIONAL CURRICULUM HSW LINK

2. *Practical and enquiry skills*
   a. plan to test a scientific idea, answer a scientific question or solve a scientific problem
   b. collect data from primary or secondary sources, including using ICT sources and tools
   d. evaluate methods of collection of data and consider their validity and reliability as evidence

RESOURCES:
Task Sheet 45, enough for one per learner.

TIME:
15 minutes for the activity, 10 minutes for class discussion.

NOTES

- This task is suitable for use as a Main or Homework Activity.
- Learners with reading difficulties will need support either by a peer or by the teacher reading through the text.
- Learners must be aware of these key words/concepts before attempting the task: planning; secondary resources; acid rain; bias.

SUGGESTED ANSWERS

The focus of this task is on the process of planning the activity. The planned activity does not need to be carried out.

- What is the question you are trying to answer? Example:
  How much acid rain has fallen in (name of town/city/county) in the last 20 years?
- What scientific evidence will you look for?
  Quantitative. Local data tables. Meteorological data and records.
- Will you use a search engine? Which key words could you try?
  Name of search engine, with reason. Key words could be anything relevant, e.g. acid rain + (place name), pollution + (place name), etc.
- Will you use the library? Which resources will you use?
  Newspaper records, local weather station data, etc.
- How will you ensure your data is unbiased?
  Know the source of the data, e.g. environmental group publications may exaggerate claims, meteorological office should be unbiased, etc.

EXTENSION SUGGESTION

Try the search.

ACID RAIN RESEARCH

Has there been acid rain in your area over the past 20 years?

How could you find out using secondary resources?

TASK

Plan to use secondary resources to find out if there has been acid rain in your local area over the past 20 years.

Write down the steps you intend to take to gather scientific, reliable and accurate data.

Include:
- What is the question you are trying to answer?
- What scientific evidence will you look for?
- Will you use a search engine? Which key words could you try?
- Will you use the library? Which resources will you use?
- How will you ensure your data is unbiased?

PLANETARY ATMOSPHERES

46

KS4 NATIONAL CURRICULUM HSW LINK

3. Communication skills
   a. recall, analyse, interpret, apply and question scientific information or ideas
   b. use both qualitative and quantitative approaches
   c. present information, develop an argument and draw a conclusion, using scientific, technical and mathematical language, conventions and symbols, and ICT tools

RESOURCES:
Task Sheet 46, enough for one per learner. May also need graph paper.

TIME:
15 minutes for the activity, 5 minutes for class discussion.

NOTES

- This task is most suitable for use as a Main or Homework Activity.
- Learners must be aware of these key words/concepts before attempting the task: graph drawing; atmosphere.

SUGGESTED ANSWERS

A. What sort of graph will you draw?
   Bar graph
   How will you label your *x*-axis?
   Planet
   What scale and labels will you use on the *y*-axis?
   Percentage of gas in atmosphere (%)
B. Which style of presentation, the table or your graph, is the best way to show the data? *The graph is the best way to present the data because it is clear. It is easy to see differences and similarities in the gases and the amount of gases present in the atmospheres of the three planets. (However, it is difficult to distinguish the amounts of minor gases.)*
C. Can you suggest another way of presenting the data?
   The data could have been presented as a set of three pie charts, with the minor gases plotted on their own pie chart.

EXTENSION SUGGESTION

Find out about the atmospheres of other planets in the solar system.

PLANETARY ATMOSPHERES

46

The table below shows the composition of the atmospheres of the planets Venus, Earth and Mars:

Planet	Content of gases in atmosphere (%)					
	Carbon dioxide	Nitrogen	Oxygen	Argon	Water vapour	Carbon monoxide
Venus	96.5	3.5				
Earth	0.04	78	21	0.1	1	
Mars	95.3	2.7	0.13	1.6	0.03	0.07

TASK

A. On the grid above, draw a graph to represent this information.
    –   What sort of graph will you draw?
    –   How will you label your x-axis?
    –   What scale and labels will you use on the y-axis?

B. Which style of presentation, the table or your graph, is the best way to show the data?

C. Can you suggest another way of presenting the data?

KS4 NATIONAL CURRICULUM HSW LINK

3. *Communication skills*
   a. recall, analyse, interpret, apply and question scientific information or ideas

RESOURCES:
Task Sheet 47, cut up into 14 cards (1 instruction and 13 sorting cards), one set between two to four learners.

TIME:
15 minutes for the activity, 5 minutes for class discussion.

NOTES

- This task is most suitable for use as an extended Starter, Main or Homework Activity.
- Learners must be aware of these key words/concepts before attempting the task: Haber process.

SUGGESTED ANSWERS

Nitrogen molecules from the air are drawn into a reaction vessel that is filled with hydrogen molecules.
Inside the vessel, the nitrogen molecules find themselves under a high pressure of 200 atmospheres that…
pushes the nitrogen and hydrogen molecules closer together and a temperature of 400°C that…
makes the nitrogen and hydrogen molecules move around faster.
With an iron catalyst to help, these conditions make it much easier for the nitrogen molecules to react with the hydrogen molecules, forming…
liquid ammonia, which is made up of…
one nitrogen atom and three hydrogen atoms.
This is a reversible reaction, so the nitrogen and hydrogen atoms move backwards and forwards between being gas nitrogen molecules and an ammonia molecule.
The liquid ammonia is drained away from the reaction vessel and the unreacted gases are…
recycled back into the reaction vessel. The liquid ammonia is then taken to another…
…reaction vessel, where the ammonia is reacted with sulphuric acid.
The reaction forms the salt, ammonium sulphate, that can be used…
on fields as fertiliser, where plants can absorb the nitrogen to make proteins and grow.

EXTENSION SUGGESTION

Write the word equations at relevant stages.

FERTILISER FROM THE AIR

TASK

Sort the cards into the order that tells the story of how nitrogen molecules from the air become artificial fertilisers that help plants to grow.
Inside the vessel, the nitrogen molecules find themselves under a high pressure of 200 atmospheres that...
recycled back into the reaction vessel. The liquid ammonia is then taken to another...
liquid ammonia, which is made up of...
...reaction vessel, where the ammonia is reacted with sulphuric acid.
The reaction forms the salt, ammonium sulphate, that can be used...
on fields as fertiliser, where plants can absorb the nitrogen to make proteins and grow.
pushes the nitrogen and hydrogen molecules closer together and a temperature of 400°C that...
This is a reversible reaction, so the nitrogen and hydrogen atoms move backwards and forwards between being gas nitrogen molecules and an ammonia molecule.
The liquid ammonia is drained away from the reaction vessel and the unreacted gases are...
makes the nitrogen and hydrogen molecules move around faster.
With an iron catalyst to help, these conditions make it much easier for the nitrogen molecules to react with the hydrogen molecules, forming...
one nitrogen atom and three hydrogen atoms.
Nitrogen molecules from the air are drawn into a reaction vessel that is filled with hydrogen molecules.

IS THE OZONE LAYER RECOVERING?

48

KS4 NATIONAL CURRICULUM HSW LINK

3. *Communication skills*
   a. recall, analyse, interpret, apply and question scientific information or ideas
   c. present information, develop an argument and draw a conclusion, using scientific, technical and mathematical language, conventions and symbols, and ICT tools

RESOURCES:
Task Sheet 48, enough for one per learner. Graph paper.

TIME:
15 minutes for the activity, 5 minutes for class discussion.

NOTES

- This task is most suitable for use as an extended Starter, or Main or Homework Activity.
- Learners must be aware of these key words/concepts before attempting the task: graph drawing; graph interpretation; ozone.

SUGGESTED ANSWERS

A. Identify the independent and dependent variables.
   Independent variable – year; Dependent variable - amount of ozone (DU).
B. Plot a graph using this data.
   Students should plot a line graph with a line of best fit.
C. Some scientists have said that the reduction in the use of CFCs has led to the first steps in the recovery of the ozone layer. Does your graph support this? What further evidence would you need to support this argument?
   Although the line levels begin to go up at the end, most scientists agree it is too soon to say that a definite recovery has happened and that we need to wait another 10 years.

EXTENSION SUGGESTION

Suggest why old fridges and freezers have to be disposed of carefully.

IS THE OZONE LAYER RECOVERING?

48

The ozone layer is a layer of O_3 molecules between 20 and 50 kilometres above the Earth's surface. The ozone layer protects the Earth from ultra-violet radiation. Scientists observed that the ozone layer above Antarctica was thinning and being destroyed by chemicals containing the halogens chlorine, bromine and fluorine. One example of an ozone destroying chemical was chlorofluorocarbon (CFC), used as aerosol propellant and coolant in refrigeration units.

In 1987, an international agreement was signed by politicians from around the world restricting the use of chemicals such as CFCs which destroy the ozone layer.

This table shows the amount of ozone in the atmosphere since 1980.
[The ozone layer is measured in Dobson units (DU).]

Year	Amount of Ozone (DU)
1980	194
1982	195
1984	154
1986	140
1988	109
1990	108
1992	94
1994	105
1996	100
1998	83
2000	94
2002	92
2003	120
2004	92

TASK

A. Identify the independent and dependent variables.

B. Plot a graph using this data.

C. Some scientists have said that the reduction in the use of CFCs has led to the first steps in the recovery of the ozone layer. Does your graph support this? What further evidence would you need to support this argument?

COAL AS AN ENERGY RESOURCE: BENEFITS, DRAWBACKS AND RISKS

49

KS4 NATIONAL CURRICULUM HSW LINK

4. *Applications and implications of science*
   a. about the use of contemporary scientific and technological developments and their benefits, drawbacks and risks

RESOURCES:
Task Sheet 49, cut into 9 cards (1 instruction and 8 sorting cards), one between two or four learners.

TIME:
10 minutes for the activity, 10 minutes for class discussion.

NOTES

- This task is suitable for use as a Starter, Main Activity or Plenary.
- Learners must be aware of these key words/concepts before attempting the task: benefit, drawback and risk.

SUGGESTED ANSWERS

Below is the intention of each statement: Benefit (B), Drawback (D) or Risk (R).

- When coal is burned, mercury is released into the air and settles in water. Mercury can accumulate in fish and shellfish, and can harm animals, including humans, that eat them. *(R)*
- Coal is non-renewable, when it is gone it can't be replaced. *(D)*
- Burning coal releases sulphur dioxide, which reacts with water in the atmosphere to form acid rain. *(R)*
- Coal is a reliable source of energy. *(B)*
- There are large reserves of coal distributed around the Earth. *(B)*
- Coal is cheap. *(B)*
- Burning coal releases nitrogen oxides, which cause smog and contribute to acid rain. *(R)*
- Coal releases vast amounts of carbon dioxide when it is burned as a fuel. Carbon dioxide contributes towards global warming. *(R)*

EXTENSION SUGGESTION

List any other benefits, drawbacks or risks you can think of that relate to coal being used as an energy resource.

COAL AS AN ENERGY RESOURCE: BENEFITS, DRAWBACKS AND RISKS

TASK

Discuss each statement card and decide whether it is a benefit, drawback or a risk of using coal as an energy resource.

- A **benefit** is something that generally has a good effect on people.
- A **drawback** is something that is a hindrance or is the 'downside'.
- A **risk** is a possible danger or source of harm.

When coal is burned, mercury is released into the air and settles in water. Mercury can accumulate in fish and shellfish, and can harm animals, including humans, that eat them.	Coal is non-renewable, when it is gone it can't be replaced.
Burning coal releases nitrogen oxides, which cause smog and contribute to acid rain.	Coal is a reliable source of energy.
There are large reserves of coal distributed around the Earth.	Coal is cheap.
Burning coal releases sulphur dioxide, which reacts with water in the atmosphere to form acid rain.	Coal releases vast amounts of carbon dioxide when it is burned as a fuel. Carbon dioxide contributes towards global warming.

GROWING BIOFUELS IN MALAWI: SOCIAL, ECONOMIC AND ENVIRONMENTAL ISSUES

50

KS4 NATIONAL CURRICULUM HSW LINK

4. *Applications and implications of science*
   b. to consider how and why decisions about science and technology are made, including those that raise ethical issues, and about the social, economic and environmental effects of such decisions

RESOURCES:
Task Sheet 50, cut up into 7 cards (1 instruction and 6 sorting cards), one set between two to four learners.

TIME:
10 minutes for the activity, 10 minutes for class discussion.

NOTES

- This task is most suitable for use as a Starter, Main Activity or Plenary.
- Learners must be aware of these key words/concepts before attempting the task: social, economic and environmental; multinational; maize.

SUGGESTED ANSWERS

Below is the intention of each statement: Social (S), Economic (Ec) or Environmental (Env).
- "With increasing oil prices, biofuels provide a green, affordable alternative to diesel and petrol." **Daniel, Multinational Company CEO** *(Ec/Env)*
- "Without strict regulation, forests may be cut down so the land can be used for planting biofuels. This will result in a loss of biodiversity and possibly increased carbon dioxide emissions." *(Env)*
  Bernard, Environmental Campaigner
- "We are under pressure to sell our farming land to multinational companies so they can grow biofuels on our land." *(S)*
  Mwaba, Subsistence Farmer
- "If crops, like corn, can be used as biofuels, their cost will go up – making them less affordable to local people." *(Ec)*
  Maggie, Local Politician
- "Is it right that the maize that would feed one person for one year should be used to make enough fuel to fill the tank of a 4x4 once?" *(S)*
  Monica, Human Rights Campaigner
- "Biofuels are renewable energy resources which could reduce carbon dioxide by up to 60% compared with fossil fuels." *(Env)*
  Adam, British Government Official

EXTENSION SUGGESTION

Suggest some rules which would prevent environmental damage when growing biofuels.

Growing Biofuels in Malawi: Social, Economic and Environmental Issues

50

TASK

Discuss each statement card and decide whether it is a social, economic or environmental issue about growing biofuels in the African country of Malawi.

- **Social**: To do with people's lives and the effect on running a society.
- **Economic**: To do with money, either making money or keeping costs down.
- **Environmental**: To do with keeping our environment unpolluted.

"With increasing oil prices, biofuels provide a green, affordable alternative to diesel and petrol." **Daniel, Multinational Company CEO**	"Without strict regulation, forests may be cut down so the land can be used for planting biofuels. This will result in a loss of biodiversity and possibly increased carbon dioxide emissions." **Bernard, Environmental Campaigner**
"We are under pressure to sell our farming land to multinational companies so they can grow biofuels on our land." **Mwaba, Subsistence Farmer**	"If crops, like corn, can be used as biofuels, their cost will go up - making them less affordable to local people." **Maggie, Local Politician**
"Is it right that the maize that would feed one person for one year should be used to make enough fuel to fill the tank of a 4x4 once?" **Monica, Human Rights Campaigner**	"Biofuels are renewable energy resources which could reduce carbon dioxide by up to 60% compared with fossil fuels." **Adam, British Government Official**

Badger Publishing Limited
15 Wedgwood Gate
Pin Green Industrial Estate
Stevenage, Hertfordshire SG1 4SU
Telephone: 01438 356907
Fax: 01438 747015
www.badger-publishing.co.uk
enquiries@badger-publishing.co.uk

Badger GCSE Science
How Science Works
Chemistry

First published 2008
ISBN 978 1 84691 309 9

Publisher: David Jamieson
Editor: Paul Martin
Designer: Adam Wilmott
Illustrator: John Dillow (Beehive Illustration), Adam Wilmott, Juliet Breese

Cover photo: Development researcher with micropipettor © PHOTOTAKE Inc. / Alamy.

Printed in the UK

BADGER SCIENCE

NEW FOR 2008

By Andrew Grevatt

Badger *Level-Assessed Tasks* Concepts

Y7 Concepts Teacher Book and CD	ISBN 978-1-84691-301-3
Y8 Concepts Teacher Book and CD	ISBN 978-1-84691-302-0
Y9 Concepts Teacher Book and CD	ISBN 978-1-84691-303-7

Badger *Level-Assessed Tasks* HSW

Y7 HSW Teacher Book and CD	ISBN 978-1-84691-304-4
Y8 HSW Teacher Book and CD	ISBN 978-1-84691-305-1
Y9 HSW Teacher Book and CD	ISBN 978-1-84691-306-8

How Science Works activities for KS4

with Dr. Deborah Shah-Smith

Biology Teacher Book and CD	ISBN 978-1-84691-307-5
Chemistry Teacher Book and CD	ISBN 978-1-84691-309-9
Physics Teacher Book and CD	ISBN 978-1-84691-308-2

Also available from Badger Publishing – see our website for more details:

Badger *Assessment for Learning Tasks*

Andrew Grevatt and Anne Penfold

GCSE Core Topics Teacher Book and CD	ISBN 978-1-84424-952-7
Additional Topics Teacher Book and CD	ISBN 978-1-84691-207-8

Ideas About Science for KS4

Dr Mark Evans and Chris Jennings

HSW resources, including material for your VLE

Teacher Book and CD	ISBN 978-1-84691-447-8

Badger *Science Starters*

John Parker

Y7 Teacher Book	ISBN 978-1-85880-353-1
Y7 PDF CD	ISBN 978-1-84424-130-9
Y8 Teacher Book	ISBN 978-1-85880-354-8
Y8 PDF CD	ISBN 978-1-84424-131-6
Y9 Teacher Book	ISBN 978-1-85880-355-5
Y9 PDF CD	ISBN 978-1-84424-132-3

For the Interactive Whiteboard Mary Mather

Y7 SMART™	ISBN 978-1-84424-779-0
Y7 Promethean	ISBN 978-1-84424-780-6

For details of the full range of books and resources from

Badger Publishing

including

- **Book Boxes** for 11-16 and Special Needs and Class sets of novels
- **KS3 Guided Reading** – Teacher Files and book packs
- **Between the Lines** – course exploring text types at KS3
- **Under the Skin** – progressive plays for KS3
- **Full Flight, Dark Flight, First Flight & Rex Jones** for reluctant readers
- **Brainwaves** – non-fiction to get your brain buzzing
- **SAT Attack** – Badger English and Maths Test Revision Guides
- **Badger KS3 Starters** for *Literacy*, *Maths* and *Science* and for the *Foundation Subjects* – History, Geography, Religious Education, Music, Design & Technology, Modern Foreign Languages
- **Main Activity: Problem Solved!** KS3 Maths problem solving course
- **Concepts** and **How Science Works** Level-Assessed Tasks for KS3 Science
- **Badger ICT** – lesson plans for KS3
- **Badger Music** – lesson plans for KS3
- **Building Blocks History** – complete unit for KS3
- **Thinking Together in Geography** – developing thinking skills
- **Multiple Learning Activities** – providing for different learning preferences
- **Beliefs and Issues** – Badger KS3 Religious Education course
- **Black** and **Asian Pride** and **British Role Model** poster sets
- **Dual Language** readers and **Full Flight Runway** for EAL

KEY STAGE 4

- **Badger KS4 Starters** for *Maths*
- **Science Assessment for Learning Tasks** for KS4
- **How Science Works** tasks for KS4
- **Badger GCSE Religious Studies** – illustrated text books
- **Surviving Citizenship @ KS4** – teaching guide

INTERACTIVE WHITEBOARD

- **Badger IAW Activities** for KS3 Music
- **Y7 Science Starters** for Smart™ Board and Promethean
- **Full Flight Guided Writing CD** – writing activities in Word
- PDF CD versions of many titles also now available.

See our full colour catalogue, visit our website or our showroom for more:
www.badger-publishing.co.uk

Badger Publishing Limited
15 Wedgwood Gate, Pin Green Industrial Estate,
Stevenage, Hertfordshire SG1 4SU
Telephone: 01438 356907   Fax: 01438 747015
enquiries@badger-publishing.co.uk